Integrated Electronics

Rodney Bruce Sorkin

U. S. Department of Defense

McGRAW-HILL BOOK COMPANY

*New York   St. Louis   San Francisco   Dusseldorf   London
Mexico   Panama   Sydney   Toronto*

INTEGRATED ELECTRONICS

To my Mother and Father

Preface

The purpose of this book is to provide an introduction to the semiconductor and design theory necessary for understanding and using semiconductor integrated circuits. It is intended as a broad and interpretive guidebook to the physical basis, fabrication arts, and applications engineering underlying the integrated electronics technology. The reader will appreciate that the book can only touch upon a vast field of technology. As a consequence, topic selection and depth of treatment have been guided by their pertinence to the general overview, and in some cases, accuracy has been sacrificed to visualization. It is hoped that this exploration will equip the reader with an insight into the prominent factors, constraints and compromises that affect his own speciality, and provide him with the foundation for more extensive reading.

The text is addressed to the practicing engineer or technical specialist, and has been written with self-study specifically in mind. It should also be suitable for a three credit semester course at the senior college level. No specific background is assumed other than the basic mathematics and science commensurate with a technical degree. The text problems are not rigorous, but are included where necessary to enhance familiarity with the subject details and as a means of introducing future material.

The organization of the book follows the vertical progression from physics through construction to design. The basic physics of semiconductors and their conduction properties are reviewed in Chapter 2 and pave the way to discussion of the operating principles of semiconductor components employed in integrated circuits (Chapter 3). The theory of impurity diffusion, basic construction steps and alternative circuit structures are introduced in Chapter 4, and the outcome in terms of attainable device parameters and various practical constraints is presented in Chapter 5. Chapters 6, 7, and 8 are devoted to digital design concepts, starting with elementary combinatorial concepts, progressing through sequential design principles with concrete examples, and ending with the discussion and evaluation of the practical characteristics of formal integrated circuit logic configurations. In Chapter 9, the significant elements of linear integrated circuits and the use of circuits in design problems are discussed.

The text material is an expansion of a series of lectures given to a group of technologists, engineers, and scientists during the summer of 1966 while the author was associated with the Columbus Laboratories of the Battelle Memorial Institute. About one-half of the forty-five participants were electronic engineers, and the remainder represented equally the fields of metallurgy, chemistry, mechanical engineering, materials engineering, and physics.

The text reflects the many constructive suggestions obtained from the participants during the lecture series. The author gratefully acknowledges the support of Battelle Memorial Institute in preparing the lectures, and the the encouragement and helpful discussions of many associates there, particularly Larry Stember, Jr., Chief of the Advanced Electronics Division. The author is also grateful to many people in the academic world for the benefit of their teachings in course work and discussions, especially Marlin O. Thurston of the Ohio State University and H. C. Lin of the University of Maryland. The author would like to thank Dr. Thurston for permission to adapt one of his problems for Chapter 4.

Last, but certainly not least, the author is indebted to his wife, Linda, for typing the manuscript.

Rodney Bruce Sorkin

Contents

5. *Integrated-circuit Component and Intercomponent Characteristics* **68**

6. *Digital-design Concepts... Combinatorial-design Principles* **80**

ONE

Introduction

Evolution

There have probably been few developments in the era of modern electronics that have so excited their creators and so quickly gained loyal followers as the advent of integrated electronics. Integrated electronics, sometimes called microelectronics, is a technology of highly miniaturized electronic circuitry. However, miniaturization itself is only one of the important reasons for the phenomenal growth in importance of the technology. Integrated electronics has achieved speedy development and widespread utilization because it has been able to surmount a host of difficulties threatening to limit the continual growth in complexity of electronic systems. We find these difficulties arising in the evolution of electronic equipment, and especially military electronic equipment, since the end of World War II.

It was during the war that the significant military advantages gained by electronic means of detection, communication, and control became universally recognized. The sophisticated requirements of late-war equipment led to the use of vast amounts of electronics and spurred further important developments in the technology. Minimization of the power

1

consumption and size of electronic modules followed the trend toward greater mobility of weapons systems.

In the technologically oriented postwar period, requirements for national security led to the development of increasingly complex and sophisticated military electronic equipment. The more complex equipment became, the more electronic parts had to be combined to construct the system. In the early systems that used vacuum tubes, problems attendant with size, weight, and power consumption are obvious. More important and constraining were the related problems of maintaining the equipment in a constant state of readiness and realizing successful equipment performance throughout the duration of a mission. Meeting these reliability objectives imposed limitations on the attainable processing power of military electronic equipment.

The advent of the transistor improved the situation considerably. Transistors offered miniaturization, lower power requirements, eventual lower costs, and a hundredfold improvement in reliability. However, in the late 1950s and early 1960s, the sophistication of even transistorized electronic equipment began to exceed the capabilities of the military services to maintain the equipment. To make things more gloomy, a great exodus of trained career technicians from the military services began, as 20-year enlistments from the war expired.

At this point, the concept of the integrated circuit was introduced. Here, entire circuits would be constructed in a transistorlike fabrication. It was thought, and has since been confirmed, that the homogeneous (monolithic) nature of the circuit containing 10 to 100 component parts would result in a circuit reliability comparable to that of a single transistor. Further, the small size of the circuits would permit significant sections of an equipment to be placed on plug-in throwaway modules — advantageous for maintenance. At the end of the rainbow was the promise of lower cost through batch-processings and elimination of design and fabrication functions previously necessary at the circuit level of construction. Thus there has come to be a great emphasis on development and growth of the technology. Today, integrated circuits have met many of these expectations. The growth rate of the integrated-circuits industry since the first experimental unit was fabricated in 1958 has exceeded the early transistor growth rate. The practical application of integrated electronics may very well turn out to have been the most significant technological development of the 1960s.

Integrated Circuits and Reliability

In any given time interval, there is always a finite probability that an electronic component will fail to operate. The heater filament in a vacuum tube, the integrity of a socket pin, the connector cap on a resistor, and a solder joint all have the same annoying and unscrupulous statistical tendency to fail. Steps can be taken to minimize this tendency. It is well known that components "wear out"; therefore replace them periodically before wear-out. Take utmost care in supervising production of components, circuits, and systems. Inspect and reinspect equipment. After all possible precautions have been taken, there is still a very small statistical probability that a component will fail. By testing a large number of components for extended durations, the failure probability for each component can be determined. We can express this failure tendency in terms of an average time that it takes a random part to fail. Example mean-time-between-failures (MTBF) figures for high-grade military components and circuit parts are given below:

Component	MTBF
Electron tube	200,000 hours
Capacitor	20,000,000 hours
Resistor	100,000,000 hours
Electron tube socket	200,000,000 hours
Solder joint (+ wiring)	500,000,000 hours

Each component seems to have a long life when measured by these standards. However, the picture changes considerably when we envision large systems comprising thousands of these components. When we calculate the reliability of the entire system, we must statistically sum the failure probabilities of each component.* Let us do this for three sample systems: a hand communication set, a simple special-purpose control device, and something similar to a modern large-scale computer:

System	Tubes	Capacitors	Resistors	Sockets	Joints
Communication set	10	20	40	10	200
Control device	200	300	1,000	200	4,000
Computer	20,000	40,000	100,000	20,000	400,000

Performing the necessary calculations, we determine the MTBFs for each system as a whole:

System	MTBF, hours
Communication set	19,300
Control device	965
Computer	9.6

*The overall MTBF of an equipment comprising n components, which is considered to fail when any of its components fail, is

$$\frac{1}{(MTBF)_{overall}} = \frac{1}{(MTBF)_1} + \frac{1}{(MTBF)_2} + \cdots + \frac{1}{(MTBF)_n}$$

Thus, on the average, an electron-tube version of the modern computer would work properly less than 10 hours between repairs.

This simple example illustrates the so-called "tyranny of numbers," where the sheer magnitude of part count prevents the reliable realization of large-scale systems. The introduction of transistors (and/or semiconductor diodes) to the list of permissible component parts improves the situation in two ways: the substitution of a transistor (MTBF $\approx$ 10 million hours) for an electron tube offers a part-for-part reliability improvement, and the biasing simplicity of transistor circuits allows the number of associated passive components to be reduced. With these two improvements, a recalculation of the MTBF for the transistorized computer gives about 250 hours. In fact, the transistor made practical the design of complex equipments like the modern computer.

Let us now consider the MTBF of the computer when constructed with integrated circuits. An integrated circuit that can perform the function of 10 transistors and their associated passive component parts might have an MTBF of 5 million hours. Recalculation of the computer MTBF gives over 2,000 hours. Further, the small size of the integrated circuits allows significant segments of the computer to be built on single throwaway modules. Thus, when failure does occur, location and replacement of the faulty segment may be accomplished quickly by relatively unskilled persons.

Varieties of Integrated Circuits

Integrated electronics has been used to describe a variety of approaches to the microminiaturization of circuitry. Among these are high-density packaging of miniature conventional components, thin-film circuits, and monolithic integrated circuits. In this book, we will focus our attention on the last form mentioned — by far the most important economically, and the one with the greatest degree of technical sophistication.

High-density packaging of miniature conventional components involves the application of ordinary circuit components, e.g., resistors, diodes, transistors, in an interconnection matrix designed to minimize the total volume. There has always been a natural tendency to minimize size in electronics, and thus advances in this technology have evolved slowly, over a period of time, instead of experiencing a marked revolution in form factor. We might mention, however, a few notable landmarks in this progression. In the early 1950s, the National Bureau of Standards, under Navy sponsorship, developed an automated technique for fabricating miniature electronic circuits called "tinkertoys." The major objective of Project Tinkertoy was to fabricate compact, reliable, and inexpensive radio sonobuoys. Component parts were automatically mounted on

⅛-in.-square ceramic wafers, and the resultant "loaded" wafers were stacked one on top of the other to form an electronic circuit stage. In more or less a continuation of this same idea, in 1958 the Army Signal Corps began sponsorship of the "micromodule" program. Many of the newly developed components were utilized to form more reliable and compact circuit modules. This program was completed just a few years ago, after a number of successes in miniaturizing Army communication equipment. Concurrently with these efforts, a number of component suppliers developed pellet-sized component parts for assembly in "Swiss cheese" circuit boards.

Miniaturized packaging represented a change in degree rather than in kind, however, and offered few commensurate advantages in cost and reliability.

Around 1954, the idea of the "thin-film" circuit was evolved. Instead of using material to fabricate individual component parts and then interconnecting them by soldering, the various materials were formed on an insulating substrate to make a complete circuit. The thin-film technology associated with optical effects was very much applicable to this process. Specific techniques for fabricating these circuits vary, but the end result is a geometrical arrangement of thin films (10^2 to 10^4 Å in thickness) of various materials (Nichrome, aluminum, silicon monoxide, tantalum) on a ceramic or glass substrate to form the passive elements (resistors and capacitors) and interconnection leads. By taking advantage of the properties of very thin films for resistive and dielectric layers, entire circuits can be fabricated in a relatively small total area — a square inch or less — and, in most cases, can be made equivalent in performance to the conventional component-part versions.

Although interesting advances in thin-film technology continue in the present, thin-film circuits are not economically competitive with monolithic integrated circuits.

Monolithic integrated circuits arose from advances in transistor-fabrication technology. We will have occasion to discuss the properties and construction of transistors in more detail. At this point, we can say that improvements in the performance characteristics of transistors required more and more precise geometry control. A transistor is constructed of a group IV semiconductor with three adjacent areas of different impurity content. The geometry and impurity density are quite important to the transistor's characteristics. A number of techniques were developed to introduce the impurities. Impurities can be alloyed into the semiconductor, added to the original melt, or diffused into the single-crystal semiconductor at high temperatures. The latter process offered best results, but the difficulty of geometry control prevented further advances. In the late 1950s, the planar-diffused process was devised. At temperatures where impurities readily diffuse into silicon, silicon dioxide remains almost impenetrable. Thus, oxidation of the silicon can block impurities from diffusing into it. Using precise photolithographic techniques, small "windows" can be opened in the oxide to permit selective diffusion.

Technologists who had become quite proficient with diffusion techniques became interested in the idea of treating semiconductor material in such a way as to make a complete circuit. Certainly, transistors and diodes could be formed in the usual way. Passive device areas could be formed by diffusing into the semiconductor conductive paths (resistors) and by oxidizing the semiconductor to form dielectrics (capacitors). Moreover, the size of the entire circuit would not be too much greater than that of the transistor itself. Material interfaces could be eliminated to a great extent, as well as soldered connections, thus introducing orders of reliability improvement. In 1958, the first integrated circuit of this type was constructed.

As we shall see, monolithic integrated circuits are "batch-processed," i.e., a great many circuits are processed alike at one time during manufacture. A matrix of circuits is formed, one next to the other, within a circular disk or "wafer" of silicon of $1\frac{1}{4}$-in. diameter or larger. Economies of circuit fabrication dictate the minimization of individual circuit size to permit the formation of a greater number of circuits per wafer. Thus the actual active-circuit areas have tended to decrease as much as improvements in the technology would allow.

Since the packaging cost is a fairly constant cost parameter, further economies were obtained by encapsulating more than one circuit function in a single package. These considerations led to the concept of "large-scale integration," where subsystem assemblies rather than circuits have become the target of fabrication efforts. The development of a practical manufacturing technique for the field-effect transistor in the mid-1960s has assisted the realization of such large assemblies. Semiconductor-component technologists have thus progressed from the construction of transistors to that of entire subsystems.

Interdisciplinary Nature of Integrated Electronics

Integrated electronics resulted in a horizontal integration of technical specialties. In one process, material is transformed into electronic function. Thus the combined talents of materials specialists, semiconductor physicists and chemists, and electrical designers must be employed to attain the desired product. Bringing electrical leads from the semiconductor chip to the outside world, packaging the circuits, and interconnecting packaged circuits gave new problems to metallurgists and mechanical engineers.

The trend in integrated electronics, wherein individual component parts become less identifiable and merge into the overall circuit function, approaches the realization of "molecular" electronics. Here we have the notion of the efficient utilization of material to process electrical energy such that the processing operation depends on the overall physical properties of a homogeneous medium. A few interesting circuit functions have been fabricated in semiconductor material without identifiable component parts.

REFERENCES

Integrated Circuits

1. Keonjian, Edward, ed.: *Microelectronics: Theory, Design, and Fabrication,* McGraw-Hill, 1963.
2. Khambata, Adi J.: *Introduction to Integrated Semiconductor Circuits,* John Wiley, 1963.
3. Levine, Sumner N.: *Principles of Solid-State Microelectronics,* Holt, Rinehart, and Winston, 1963.
4. Lin, H. C.: *Integrated Electronics,* Holden-Day, 1967.
5. Lynn, David K., et al., eds.: *Analysis and Design of Integrated Circuits,* McGraw-Hill, 1967.
6. Schwartz, Seymour, ed.: *Integrated Circuit Technology,* McGraw-Hill, 1967.
7. Warner, Raymond M., Jr., ed.: *Integrated Circuits, Design Principles and Fabrication,* McGraw-Hill, 1965.

TWO

Physical Basis

The design of integrated circuits is constrained by parameters different from those of conventional-circuit design practice. The constraining parameters are the limiting properties of the semiconductor material itself. One's appreciation of the technology must therefore begin with an understanding of the material phenomena responsible for external circuit characteristics. We begin our study by reviewing briefly some aspects of semiconductor-device theory essential to this understanding.

Material Physics

The Intrinsic Semiconductor. Semiconducting materials have unique properties that may be employed to advantage in constructing devices. Our study is generally confined to silicon which is used as the base material for integrated circuits. In particular, we are concerned with silicon that has been grown as a single crystal, i.e., with a regular and repetitive atomic lattice structure throughout. In the pure (intrinsic) semiconductor, the atoms (of valence 4) are bound in the lattice by means of covalent bonding. In the traditional picture shown in Fig. 2-1, each atom shares one outer-shell electron from each of its four nearest neighbors, and in turn contributes one to the sharing process. Each circle of net charge +4 represents the nucleus (14 protons and 14 neutrons) plus the 10 inner-core electrons. The material as a whole is, of course, electrically neutral.

A more convenient picture for characterizing the semiconductor is an energy description. Quantum-mechanical considerations require the outer-shell valence electrons to assume only certain restricted energy levels (energy eigenvalues). Furthermore, no more than two electrons may reside at the same energy level (Pauli exclusion principle). In a large aggregate of atoms, the allowed energies of the valence electrons are close together and form a quasi-continuous band of energy levels, the valence band. An important characteristic of the semiconductor crystal is that the number of outer-shell valence electrons is exactly sufficient to fill the valence band. When an energy band is completely filled, conduction of electrons in that band is not possible.

The energy-band principle is illustrated in Fig. 2-2. As we consider increased electron energies, we find that there is a region of forbidden

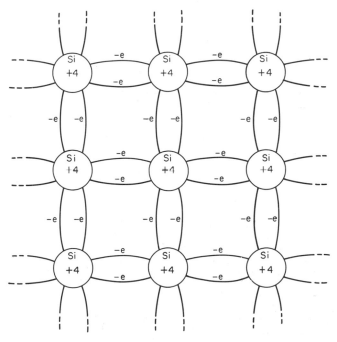

Fig. 2-1 *Schematic diagram of silicon-bound-in lattice. Each line represents an electron shared by two neighboring atoms.*

electron energies, called the forbidden-energy gap. After this energy gap, another band of allowed energies exists, the conduction band. At absolute zero temperature, all the outer-shell electrons are in the valence band, as shown in Fig. 2-2*a*. As the temperature is raised, however, thermal energy is absorbed by atoms and electrons within the lattice, and some of the outer-shell electrons will attain sufficient energy to jump to the conduction band, illustrated in Fig. 2-2*b*. In silicon, the forbidden-energy gap between the valence and conduction bands is about 1.2 ev. The ionization process is akin to a bond being broken in the picture of Fig. 2-1, with the freed electron wandering off through the crystal.

Electrons in the conduction band are free to contribute to the conduction process. Furthermore, the absence of one or more electrons in the valence band creates an equal number of positive vacancies which can be occupied by other electrons in the valence band, permitting conduction through this band. Instead of concentrating on the many-electron problem in the valence band, it has been found to be convenient to follow the motion of the few vacancies through the crystal, as shown in Fig. 2-2*c*. To these vacancies, called holes, we may ascribe particlelike properties: a positive charge of magnitude e (1.6 $\times$ 10^{-19} coulomb) and a mass comparable to that of an electron.

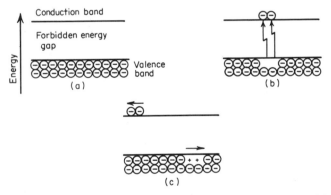

Fig. 2-2   *Energy diagram of silicon-bound-in lattice. (a) At absolute zero temperature, valence band is completely filled. (b) At finite temperature, absorption of thermal energy ionizes some electrons which jump into the conduction band. (c) Conduction is now possible in both bands. The many-electron problem in the valence band is treated by considering the movement of the positively charged vacancies.*

To summarize: the ionization of a single electron creates two particles available to the conduction process — an electron in the conduction band

and a hole in the valence band. In intrinsic material, the number of holes and electrons is equal.

Electron and Hole Densities.   From this point on, we will focus our attention on those charged particles that are available for conduction, i.e., the electrons in the conduction band and the holes in the valence band. When we talk of electrons, we shall mean free electrons in the conduction band only. As might be anticipated, the thermal generation of hole-electron pairs is matched by an opposing effect: the recombination of holes and electrons to re-form neutral atoms. From a simplified viewpoint,* the rate at which recombination occurs depends upon the probability that a hole will "meet" an electron. We might thus suspect that the recombination rate would be proportional to the product of the hole and electron densities:

$$\text{Recombination rate per unit volume} = r = \alpha n p \qquad (2\text{-}1)$$

where α = proportionality factor depending on material
n = density of electrons, electrons/cm³
p = density of holes, holes/cm³
At thermal equilibrium, the generation and recombination must be equal, and we obtain

$$g(T) = \alpha n_0 p_0 \qquad (2\text{-}2)$$

where $g(T)$ = thermal generation rate per unit volume
n_0 = equilibrium density of electrons
p_0 = equilibrium density of holes
As we discussed previously, in intrinsic material the hole and electron densities are equal: $n_0 = p_0 = n_i$. The intrinsic carrier concentration as a function of temperature may be determined from Fermi-Dirac statistics to be

$$n_i = C T^{3/2} \exp\left(-\frac{eE_g}{2kT}\right) \qquad (2\text{-}3)$$

where C = proportionality constant depending on material
T = temperature, °K
eE_g = energy gap between valence and conduction bands (1.2 ev for silicon)
k = Boltzmann's constant = 8.63×10^{-5} ev/°K
For intrinsic silicon at room temperature ($T = 300°$K),

$$n_i = 1.5 \times 10^{10}/\text{cm}^3$$

*This viewpoint, although not an accurate description of the complex mechanisms involved, is useful for getting a "feel" for the recombination process.

Let us compare the intrinsic concentration of electrons and holes with the concentration of atoms in silicon material:

$$\text{Atomic density Si} = \frac{\text{density}}{\text{atomic weight}} \times \text{Avogadro's number}$$

$$\approx 5 \times 10^{22}/\text{cm}^3$$

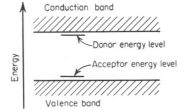

Fig. 2-3 *Donor and acceptor impurity energy levels.*

The intrinsic carrier concentration is indeed small when compared to the atomic density.

Extrinsic Semiconductors.   The substitution of impurity atoms for relatively few of the semiconductor atoms can have a significant effect on the conduction properties of the semiconductors. Impurity atoms have energy levels existing in the forbidden-energy gap, as shown in Fig. 2-3. A donor impurity is a group V element (valence of 5). Four of the outer-shell electrons are utilized in the bonding process, while the fifth is only very loosely bound (ionization energy ≈ 0.05 ev). It is thus easily ionized and provides an electron to the conduction band. However, no vacancies are created by the ionization and therefore no additional holes. The ionized atom is fixed to its lattice position and has a net electric charge of $+c$. Material doped by donor impurities is termed n-type.

An acceptor impurity is a group III element (valence of 3). With three electrons in the outer shell instead of the usual four, an electron vacancy is created. An electron from the valence band can easily jump into this vacancy and thus create a hole. An ionized acceptor atom has a net electric charge of $-c$. Material doped by acceptor impurities is termed p-type.

In most cases of interest, we may consider that the impurity atoms are 100 percent ionized, i.e., each donor (or acceptor) atom contributes one electron (or hole) to the conduction (or valence) band.

Hole and Electron Densities in Extrinsic Semiconductors.   In determining the hole and electron densities in extrinsic (doped) material, we require that any localized macroscopic region be free of net charge (space-charge neutrality). This condition implies that

$$n + N_A = p + N_D \tag{2-4}$$

where n = density of mobile electrons

p = density of mobile holes

N_A = density of ionized acceptor atoms

N_D = density of ionized donor atoms

Note that this condition is assumed to apply whether or not the carrier concentrations are at thermal equilibrium.

A second fundamental relation between hole and electron concentrations is

$$n_0 p_0 = n_i^2 \tag{2-5}$$

where the subscript 0 indicates thermal equilibrium conditions.

Relations (2-4) and (2-5) may be used to determine the equilibrium hole and electron concentrations.

As a concrete example, consider p-type material. Let n_{p_0} and p_{p_0} be the equilibrium electron and hole concentrations, respectively (the subscript p indicates p-type material). From Eq. (2-4),

$$p_{p_0} = n_{p_0} + N_A$$

Practical values of impurity concentration range from 10^{13} to 10^{20} atoms/cm³. Thus, in cases of interest, $N_A \gg n_i \gg n_{p_0}$, and

$$p_{p_0} \approx N_A \tag{2-6}$$

Nonequilibrium Carrier Concentrations. Many semiconductor devices, e.g., transistors, operate with carrier concentrations that are not at thermal equilibrium. Consider p-type material in which the carrier concentrations have been increased by some external agency. Let

$$p_p = p_{p_0} + \hat{p}_p$$
$$n_p = n_{p_0} + \hat{n}_p \tag{2-7}$$

where $\hat{p}_p$ and $\hat{n}_p$ are the excess hole and electron concentrations, respectively. The neutrality conditions of Eq. (2-4) require that

$$\hat{p}_p = \hat{n}_p \tag{2-8}$$

Since the carrier concentrations have increased, the recombination rate [Eq. (2-1)] will similarly increase. Let us determine the net increase in recombination rate due to the disturbance:

$$r = \alpha n_p p_p - \alpha n_{p_0} p_{p_0}$$
$$= \alpha[(n_{p_0} + \hat{n}_p)(p_{p_0} + \hat{p}_p) - n_{p_0} p_{p_0}]$$

$\hat{n}_p$ will ordinarily be several orders of magnitude less than p_{p_0}, so that

$$r \approx \alpha[(n_{p_0} + \hat{n}_p)p_{p_0} - n_{p_0} p_{p_0}]$$
$$\approx \alpha p_{p_0} \hat{n}_p \tag{2-9}$$
$$\approx \frac{\hat{n}_p}{\tau_n}$$

where $\tau_n = 1/\alpha p_{p_0}$ is the lifetime of electrons in the p-type material. Then, from Eq. (2-5),

$$n_{p_0} = \frac{n_i^2}{p_{p_0}} \tag{2-10}$$

In this case the holes are in the majority (majority carriers) and the electrons in the minority (minority carriers).

As an example of the use of Eq. (2-9), consider the effect of shining light on a thin slab of p-type semiconductor material. The rate of change of minority-carrier concentration is

$$\frac{dn_p}{dt} = g(T) + G - r$$

$$= g(T) + G - \alpha n_{p_0} p_{p_0} - \frac{\hat{n}_p}{\tau_n}$$

where $g(T)$ = thermal generation rate
G = generation rate of hole-electron pairs due to light
But $g(T) = \alpha n_{p_0} p_{p_0}$ from Eq. (2-2). Therefore

$$\frac{dn_p}{dt} = G - \frac{\hat{n}_p}{\tau_n} \tag{2-11}$$

After the light has been on for a long period of time, a steady-state condition is attained ($dn_p/dt = 0$). Solving for $\hat{n}_p$ we obtain

$$\Delta n_{ss} = G\tau_n \tag{2-12}$$

where Δn_{ss} is the steady-state value of excess minority-carrier concentration $\hat{n}_p$.

At $t = 0$, the light is turned off ($G = 0$). Equation (2-11) becomes

$$\frac{dn_p}{dt} = \frac{d(n_{p_0} + \hat{n}_p)}{dt} = \frac{d(\hat{n}_p)}{dt} = -\frac{\hat{n}_p}{\tau_n} \tag{2-13}$$

The solution to Eq. (2-13) is

$$\hat{n}_p = \Delta n_{ss} \, exp\left(-\frac{t}{\tau_n}\right) \tag{2-14}$$

and

$$n_p = n_{p_0} + \hat{n}_p = n_{p_0} + \Delta n_{ss} \exp\left(-\frac{t}{\tau_n}\right) \tag{2-15}$$

Note that $\hat{n}_p = \hat{p}_p$, so that

$$p_p = p_{p_0} + \Delta n_{ss} \exp\left(-\frac{t}{\tau_n}\right) \tag{2-16}$$

Analogous relations may be derived for n-type material. In particular,

$$\Delta r = \frac{\hat{p}_n}{\tau_p} \qquad (2\text{-}17)$$

where $\tau_p = 1/\alpha n_{n_0}$.

Conduction Properties of Semiconductors

Device action requires the controlled conduction of charged carriers. In a vacuum tube, the conduction is achieved by the movement of electrons in the electric field established between the plate and cathode. Control is obtained through the potential applied to the grid electrode, which modifies the electric field. In semiconductors, there are two charged carriers: negatively charged electrons and positively charged holes. There are also two mechanisms of conduction: the drift of carriers under the influence of an electric field and the diffusion of carriers from regions of high carrier concentration to regions of low concentration. We thus need to concern ourselves with four current components, namely the drift and diffusion currents of electrons and those of holes. Fortunately, many important devices may be discussed by concentrating on the most prominent of the four components: for example, the minority-carrier diffusion current in conventional bipolar transistors or the majority-carrier drift current in field-effect transistors.

Diffusion Current. Diffusion phenomena result from the tendency of particles subject to random collisions to become uniformly dispersed throughout the medium in which they are contained. The basic diffusion equation is applicable to processes in many branches of science, describing equally well the diffusion of species in gaseous mixture and the diffusion of charged carriers in semiconductors. We shall recall the diffusion equation

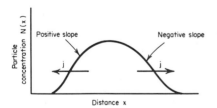

Fig. 2-4 Concentration gradient resulting in particle current flow.

later in connection with the process for introducing impurity atoms into the intrinsic semiconductor.

Consider the particle concentration diagram of Fig. 2-4. The concentration imbalance induces a flux of particles (or "particle current") to flow in such direction as to disperse the concentration imbalance. As suggested in the figure, the sign of the current is opposite to the derivative of the concentration. The current flow is given by Fick's first law:

$$f = -D \frac{\partial N}{\partial x} \tag{2-18}$$

where f = particle current density, particles/cm²-sec
$\quad$ N = concentration of particles, particles/cm³
$\quad$ D = diffusion constant relating particle-current density to negative concentration gradient, cm²/sec

In the semiconductor, each hole carries a $+e$ charge, and each electron a $-e$ charge. The respective hole and electron densities due to the diffusion are therefore

$$j_p = -eD_p \frac{\partial p}{\partial x} \tag{2-19}$$

$$j_n = eD_n \frac{\partial n}{\partial x} \tag{2-20}$$

where j_p = electric current density due to diffusion of holes, amp/cm²
$\quad$ j_n = electric current density due to diffusion of electrons, amp/cm²
$\quad$ D_p = hole diffusion constant (diffusivity), cm²/sec
$\quad$ D_n = electron diffusion constant (diffusivity), cm²/sec

In writing Eqs. (2-18) to (2-20), we have assumed one-dimensional variation. This assumption is a relatively good approximation in most devices that we will study.

Drift Current.   Charged particles, when subjected to an electric field, experience a force $F = qE$. If the particles do not interact and are free of other influences, they are accelerated with an acceleration $a = F/m = qE/m$. In a semiconductor crystal, however, collisions with surrounding atoms absorb energy from the accelerated charge carriers and limit their velocities. The result is an average "drift" of the holes and electrons along the electric field lines with a velocity magnitude proportional to the electric field strength:

$$v = \mu E \tag{2-21}$$

The constant of proportionality, μ, is called the particle mobility. By convention it is taken to be a positive quantity. Thus, a negative sign must be used in Eq. (2-21) for negatively charged particles such as electrons, which travel in a direction opposite to the electric field. The particle current density is the product of the particle concentration times the average particle drift velocity. Thus, the particle currents for holes and electrons are

$$f_p = \mu_p p E$$

$$f_n = -\mu_n n E$$

Upon multiplying the particle currents above by the charge carried by the respective particles, we have the hole and electron drift-current components:

$$j_p = e\mu_p pE \tag{2-22}$$

$$j_n = e\mu_n nE \tag{2-23}$$

where j_p = electric current density due to drift of holes, amp/cm²
j_n = electric current density due to drift of electrons, amp/cm²
μ_p = hole mobility, cm²/volt-sec
μ_n = electron mobility, cm²/volt-sec
E = electric field strength, volts/cm

Transport Equations. By adding the diffusion and drift components for holes and electrons, Eqs. (2-19), (2-20), (2-22), and (2-23), we obtain expressions for the hole and electron current densities, or the transport equations:

$$j_p = -eD_p \frac{\partial p}{\partial x} + e\mu_p pE \tag{2-24}$$

$$j_n = eD_n \frac{\partial n}{\partial x} + e\mu_n nE \tag{2-25}$$

The total electric current density is, of course, the sum of the hole and electron components. The current is then the integral of the current density over the cross-sectional area, or with the assumed one-dimensional variation,

$$I = jA \tag{2-26}$$

where I = electric current, amp
A = appropriate cross-sectional area, cm²

Continuity Conditions. The continuity conditions require that the net rate of carrier flow into an elemental volume be equal to the rate of increase of carriers within that elemental volume. The continuity conditions are applied separately to holes and electrons:

$$\frac{\partial p}{\partial t} = -\frac{1}{e}\frac{\partial j_p}{\partial x} - \frac{\hat{p}}{\tau} \tag{2-27}$$

$$\frac{\partial n}{\partial t} = \frac{1}{e}\frac{\partial j_n}{\partial x} - \frac{\hat{n}}{\tau} \tag{2-28}$$

The first term on the right of Eqs. (2-27) and (2-28) is the rate at which carriers accumulate within a microscopic region due to the differences between the current entering and leaving, and the second term is the rate at which carriers are lost through recombination.

We recall that $\hat{p} = \hat{n}$ and that the value of τ used in the above equation depends upon the material type and characteristics.

Poisson's Equation.   So far in our discussion, we have assumed space-charge neutrality over macroscopic semiconductor regions. It may happen that in certain regions (for example, around a p-n junction) a certain net space charge is established. In this case, the charge density is given by

$$\rho = (N_D - N_A + p - n)e \tag{2-29}$$

and from Poisson's equation

$$-\frac{d^2V}{dx^2} = \frac{dE}{dx} = \frac{\rho}{\epsilon} \tag{2-30}$$

an electric field will exist in the region.

Mathematical Model of Semiconductor.   Equations (2-24) to (2-30) represent a mathematical model of semiconductor material, relating conduction currents to the carrier concentrations and electric fields that exist within the material. By appropriately applying these equations according to the boundary conditions characteristic of specific devices, the characteristics and operating principles of the devices may be deduced.

Illustrative Problems

Resistivity.   Resistivity is defined as the proportionality constant between drift current density and the electric field inducing the current:*

$$E = \rho j \tag{2-31}$$

The total drift current density is the sum of hole and electron components, which is

$$j = e(\mu_p p + \mu_n n)E \tag{2-32}$$

Therefore

$$\rho = \frac{1}{e(\mu_p p + \mu_n n)} \tag{2-33}$$

Field-free Diffusion of Minority Carriers.   We will often have occasion to examine the behavior of the excess minority-carrier concentration in a homogeneous slab of semiconductor material without considering the majority carriers at all. Such analysis is possible because the excess majority-carrier concentration is exactly determined from the minority-carrier concentration by the neutrality condition. There is some interaction between the two charge carriers, but the effects can be treated by adjusting the values of the mobility and diffusion constants. The second-order corrections required are beyond the scope of our study.

---

*The symbol ρ in Eq. (2-31) for resistivity should not be confused with the charge density of Eqs. (2-29) and (2-30). Standard usage demands redundant application of this symbol; however, the meaning should be clear from the context.

Addressing our attention solely to minority carriers is desirable for the study of p-n junction devices (e.g., transistors) because the characteristics of such devices depend almost exclusively on minority-carrier flow. The following problem corresponds well to the current-flow process in the base region of transistors.

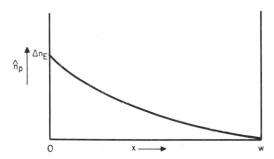

Fig. 2-5 *Field-free diffusion in slab of semiconductor material.*

Consider the slab of p-type semiconductor material of width w shown in Fig. 2-5. At $x = 0$, a constant excess minority-carrier concentration $\hat{n}_p(0) = \Delta n_E$ is maintained. At $x = w$, the excess minority-carrier concentration is constrained to zero, $\hat{n}_p(w) = 0$. We assume that the slab is field-free ($E = 0$). The problem is to determine $\hat{n}_p$ as a function of distance through the slab under steady-state conditions.

From Eqs. (2-25) and (2-28), with $E = 0$ and $\partial n/\partial t = 0$, we have

$$D_n \frac{d^2 n_p}{dx^2} = \frac{\hat{n}_p}{\tau_n} \tag{2-34}$$

or since $n_p = n_{p_0} + \hat{n}_p$,

$$D_n \frac{d^2 \hat{n}_p}{dx^2} = \frac{\hat{n}_p}{\tau_n} \tag{2-35}$$

We define L_n^2 as equal to $D_n \tau_n$. L_n has the dimensions of length and is called the diffusion length of electrons in p-type material. Equation (2-35) may be written

$$\frac{d^2 \hat{n}_p}{dx^2} = \frac{1}{L_n^2} \hat{n}_p \tag{2-36}$$

Equation (2-36) is a common differential form which has the general solution

$$\hat{n}_p = A \exp \frac{x}{L_n} + B \exp \left(-\frac{x}{L_n} \right) \tag{2-37}$$

Note that if the slab had infinite width we would have to reject the first term in the solution (it would be infinite at $x = \infty$) and we would obtain

$$\hat{n}_p = \Delta n_E \exp\left(-\frac{x}{L_n}\right) \tag{2-38}$$

where the constant $B = \Delta n_E$ is determined from the boundary condition at $x = 0$. L_n is thus identified as the distance into the material (of infinite width) for which $\hat{n}_p = \Delta n_E \exp(-1) \approx 0.37\,\Delta n_E$.

For small widths of material, we shall find it more convenient to use the equivalent expression for Eq. (2-37) in terms of hyperbolic functions:

$$\hat{n}_p = C \cosh\frac{x}{L_n} + D \sinh\frac{x}{L_n} \tag{2-39}$$

where $\cosh y \equiv [\exp y + \exp(-y)]/2$
$\qquad \sinh y \equiv [\exp y - \exp(-y)]/2$

We determine the constants of the solution from the boundary conditions.

At $x = 0$, $n_p = \Delta n_E$ and therefore $C = \Delta n_E$. At $x = w$, $\hat{n}_p = 0$. This condition requires that

$$D = \Delta n_E \frac{\cosh\,(w/L_n)}{\sinh\,(w/L_n)} = \Delta n_E \coth\frac{w}{L_n}$$

Thus the complete solution to our problem is

$$\hat{n}_p = \Delta n_E \left(\cosh\frac{x}{L_n} - \coth\frac{w}{L_n}\sinh\frac{x}{L_n}\right) \tag{2-40}$$

In transistors the width of the base region is made small with respect to L_n ($w \ll L_n$). Under these conditions Eq. (2-40) becomes almost linear, as may be demonstrated by using product expansions for the hyperbolic functions.

$$\sinh y = y + \frac{y^3}{3!} + \frac{y^5}{5!} + \cdots$$

$$\cosh y = 1 + \frac{y^2}{2!} + \frac{y^4}{4!} + \cdots$$

If we treat as negligible all terms of second order or greater, we obtain

$$\hat{n}_p = \Delta n_E \left(1 - \frac{x}{w}\right) \tag{2-41}$$

Lumped Model of Semiconductor. Equations (2-24), (2-25), (2-27), and (2-28) are differential equations in space and time derivatives. As the second illustrative problem indicates, it is often possible to simplify equations when small dimensions are involved (e.g., base region of transistor). One approach to simplification is to eliminate the space derivatives from the equations by approximating the derivatives by finite difference "jumps." Thus $\partial p/\partial x$ becomes $(p_2 - p_1)/(x_2 - x_1)$, etc. This procedure allows lumped constants to be defined, similar to electrical-component constants (resistance, capacitance, etc.) that bear a one-to-one correspondence to physical processes in the semiconductor.

As an example of this approach, consider the field-free diffusion of minority carriers in n-type material. We start by integrating Eq. (2-27) over an incremental volume $A\,dx$. For simplicity, we dispense with the subscript n on the concentration variables that would normally be used to indicate the n-type material:

$$eA \int_x^{x+\Delta x} \frac{\partial p}{\partial t}\,dx = -A \int_x^{x+\Delta x} \frac{\partial j_p}{\partial x}\,dx - \frac{eA}{\tau_p} \int_x^{x+\Delta x} \hat{p}\,dx$$

or

$$I_p(x) - I_p(x + \Delta x) = (eA\,\Delta x)\frac{\partial \hat{p}}{\partial t} + \left(\frac{eA\,\Delta x}{\tau_p}\right)\hat{p} \tag{2-42}$$

where we have made use of the fact that $\partial p/\partial t - \partial(p_0 + \hat{p})/\partial t - \partial\hat{p}/\partial t$, since p_0, the equilibrium concentration, is assumed to be constant.

If we confine the current flow to field-free diffusion ($E = 0$), then we have from Eq. (2-24)

$$I_p = -eA\,D_p\frac{\partial p}{\partial x} = -eA\,D_p\frac{\partial \hat{p}}{\partial x} \quad = \quad -qAD_p\left(\frac{\hat{p}_2 - \hat{p}_1}{\Delta x}\right)$$

Consider the two adjacent elemental volumes of Fig. 2-6, with given input and output currents. In the development of the model, the small incremental volumes are assumed to have discrete values of excess minority-carrier concentration, $\hat{p}_1$ and $\hat{p}_2$. The derivative is approximated by

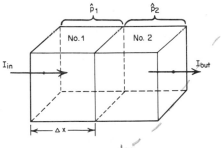

Fig. 2-6   *Elemental volumes in semiconductor treated as discrete "lumps."*

$\partial p/\partial x = (\hat{p}_2 - \hat{p}_1)/\Delta x$. Thus, writing equations for the volumes 1 and 2 in Fig. 2-6, we have

$$I_{in} = (eA\,\Delta x)\frac{d\hat{p}_1}{dt} + \left(\frac{eA\,D_p}{\Delta x}\right)(\hat{p}_1 - \hat{p}_2) + \left(\frac{eA\,\Delta x}{\tau_p}\right)\hat{p}_1 \tag{2-43}$$

$$I_{out} = -(eA\,\Delta x)\frac{d\hat{p}_2}{dt} + \left(\frac{eA\,D_p}{\Delta x}\right)(\hat{p}_1 - \hat{p}_2) - \left(\frac{eA\,\Delta x}{\tau_p}\right)\hat{p}_2 \tag{2-44}$$

It is noted that each term in Eq. (2-43) or (2-44) represents a current or equivalent current necessary to satisfy continuity conditions. For example, in Eq. (2-43), the term on the left is an input hole current entering the incremental volume. The first term on the right side of the equation is that part of the input current being stored through an increasing concentration. The second term is the amount of current diffusing from the first volume to the second. The third term is the current lost through recombination.

The coefficients of the charge-carrier variables in Eqs. (2-43) and (2-44) may be associated with lumped parameters that are functions of geometry and physical constants. These are called the Linvill parameters* and are identified as follows:

Storance: $\quad S = eA\ \Delta x$

Diffusance: $\quad H_D = eA\ \dfrac{D_p}{\Delta x}$

Combinance: $\quad H_C = eA\ \dfrac{\Delta x}{\tau_p}$

The use of these parameters will be clarified when we discuss actual devices. The important point to remember is that we have eliminated space derivatives by assuming small discrete regions, and that the current entering a given volume can be related to the carrier concentration by means of lumped parameters.

p-n Junction Characteristics

Einstein Relation.   Both the mobility and diffusion constants are a measure of the relative ease with which mobile charge carriers are transported through semiconductor material. It is therefore not surprising that the two constants are related in a particularly simple way. This relationship, called the Einstein relation, is

$$\frac{D_n}{\mu_n} = \frac{D_p}{\mu_p} = \frac{kT}{e} \tag{2-45}$$

The proof of the Einstein relation is beyond the scope of our study.

The p-n Junction.   A p-n junction is the boundary joining a p-type region and an n-type region in an otherwise homogeneous semiconductor. An intuitive picture of the equilibrium charge distribution around the

*For further treatment of lumped parameters see John G. Linvill, *Models of Transistors and Diodes*, McGraw-Hill, 1963.

junction may be obtained from Fig. 2-7. The p-type material is doped with an acceptor density N_A and the n-type region with a donor density N_D.

If we imagine the two differently doped types just joined together, the charge distribution appears as shown in Fig. 2-7a. The large circles represent the immobile ionized acceptor and donor impurity ions, and the small plus and minus signs the mobile hole and electron charge carriers. Each region is electrically neutral and no electric field exists across the junction. We know, however, that this condition cannot be long-lasting. The holes will tend to diffuse to the right and the electrons to the left. When these particles recombine, they will leave a net unneutralized charge behind — the immobile impurity ions. An electric field will be established that will tend to retard the further diffusion of charge carriers. At equilibrium, as shown in Fig. 2-7b, the electric field will just be sufficient to prevent further diffusion, and the hole and electron currents will be individually equal to zero. Since we do not expect the electric field to extend into the remaining neutral regions, the number of uncompensated acceptor ions (negative charge) must equal the number of uncompensated donor ions (positive charge). The region devoid of carriers is called the depletion region.

"Built-in" Junction Voltage. The electric field existing across a p-n junction results in a potential difference between the two regions. We

p-type (N_A)   n-type (N_D)

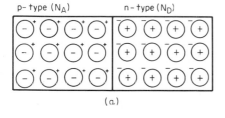

(a)

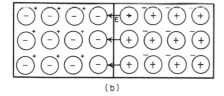

(b)

Fig. 2-7 *p-n junction* **(a)** *prior to and* **(b)** *at equilibrium.*

are now in a position to calculate the magnitude of this potential, called the "built-in" junction voltage.

We know that both the hole and electron currents are zero at equilibrium. From Eq. (2-25), with $j_n = 0$, we obtain

$$0 = eD_n \frac{dn}{dx} + e\mu_n nE \qquad (2\text{-}46)$$

Use of the Einstein relation, Eq. (2-45), gives

$$\frac{dn}{dx} + \frac{eE}{kT} n = 0 \tag{2-47}$$

We may solve Eq. (2-47) for the electric field:

$$E = -\frac{kT}{e} \frac{1}{n} \frac{dn}{dx} \tag{2-48}$$

Now

$$V_B = -\int_{p \text{ side}}^{n \text{ side}} E \, dx = \frac{kT}{e} \ln n \Big|_{p \text{ side}}^{n \text{ side}}$$

$$= \frac{kT}{e} \ln \left(\frac{n_{n_0}}{n_{p_0}} \right) \tag{2-49}$$

where n_{n_0} = thermal equilibrium value of n in n region
n_{p_0} = thermal equilibrium value of n in p region
V_B = built-in junction voltage
Since $n_{n_0} \approx N_D$ and $n_{p_0} \approx n_i^2/N_A$, Eq. (2-49) may be rewritten as

$$V_B = \frac{kT}{e} \ln \frac{N_A N_D}{n_i^2} \tag{2-50}$$

Had we used relation (2-24) instead of (2-25), we would have obtained

$$E = \frac{kT}{e} \frac{1}{p} \frac{dp}{dx} \tag{2-51}$$

and

$$V_B = -\frac{kT}{e} \ln \frac{p_{n_0}}{p_{p_0}} = \frac{kT}{e} \ln \frac{p_{p_0}}{p_{n_0}} \tag{2-52}$$

which gives exactly Eq. (2-50).

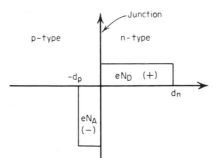

Fig. 2-8 Charge profile of region surrounding p-n junction.

Width of Depletion Region. The depletion region extends into both material types to an extent dictated by the electric field requirement. This region of uncompensated charge may be depicted as shown in Fig. 2-8. The distances into the p- and n-type materials we represent by d_p and d_n, respectively. In later discussion of integrated-circuit capacitors, we will need to know the values of these distances. Since the negative charge on the p side of the junction must equal the positive charge on the n side, we have

$$eAN_D\, d_n = eAN_A\, d_p \tag{2-53}$$

Thus the areas on both sides of the junction in Fig. 2-8 must be equal. To determine the required distances, we apply Poisson's equation:

$$\frac{dE}{dx} = \frac{\rho}{\epsilon} \tag{2-54}$$

where ρ is the charge density given by $-eN_A$ on the p side, $e\,N_D$ on the n side.

In the p region,

$$E = \int - \frac{eN_A}{\epsilon}\, dx = -\frac{eN_A}{\epsilon}\, x + C_1 \tag{2-55}$$

The constant of integration is determined by the condition

$$E = 0 \qquad \text{at } x = -d_p$$

which gives

$$E = -\frac{eN_A}{\epsilon}\, (x + d_p) \tag{2-56}$$

Similarly for the n region we obtain

$$E = +\frac{eN_D}{\epsilon}\, (x - d_n) \tag{2-57}$$

Let us assume that we have applied a voltage V_a (in forward direction with most positive potential on the p side) in addition to the "built-in" junction voltage V_B (in reverse direction). The total voltage rise is

$$V_t = V_R - V_a \tag{2-58}$$

The electric field is the negative gradient of potential or

$$E = -\frac{dV}{dx} \tag{2-59}$$

We may integrate Eq. (2-59) over the entire depletion region to determine the total voltage V_t:

$$
\begin{aligned}
V_t &= -\int_{-d_p}^{d_n} E\, dx \\
&= \int_{-d_p}^{0} \frac{eN_A}{\epsilon}\, (x + d_p) + \int_{0}^{d_n} -\frac{eN_D}{\epsilon}\, (x - d_n) \\
&= \frac{e}{2\epsilon}\, (N_A\, d_p^2 + N_D\, d_n^2) \tag{2-60}
\end{aligned}
$$

Equations (2-53) and (2-60) may be solved simultaneously for d_p and d_n to obtain

$$d_p = \left(\frac{2\epsilon V_t}{e} \frac{N_D}{N_A N_D + N_A^2} \right)^{1/2}$$

$$d_n = \left(\frac{2\epsilon V_t}{e} \frac{N_A}{N_A N_D + N_D^2} \right)^{1/2}$$

(2-61)

The total depletion width is therefore

$$d = d_n + d_p = \left[\frac{2\epsilon V_t}{e(N_A + N_D)} \right]^{1/2} \left[\left(\frac{N_A}{N_D} \right)^{1/2} + \left(\frac{N_D}{N_A} \right)^{1/2} \right]$$

(2-62)

There are several important points to emphasize with respect to Eqs. (2-61) and (2-62):

1. An applied voltage in the forward direction will decrease the depletion width. An applied voltage in the reverse direction will increase the depletion width.

2. The relative extent of the depletion region on either side of the junction is inversely proportional to the impurity-concentration ratios ($d_n/d_p = N_A/N_D$).

3. If one side is much more heavily doped than the other, we may neglect the depletion width on that side. For $N_A \gg N_D$,

$$d \approx d_n \approx \left(\frac{2\epsilon V_t}{eN_D} \right)^{1/2}$$

(2-63)

Nonequilibrium Conditions: Minority-carrier Injection. Equations (2-50) and (2-52) were derived under the assumption of equilibrium conditions. The hole and electron currents, zero under equilibrium, are determined by the difference between large drift and diffusion components in balance. Under nonequilibrium conditions with an applied voltage that supports a net current flow, we may anticipate that the previous development will remain valid provided that we restrict ourselves to small currents. Operation under this restriction is called the "low-injection regime." Our approximation yields the equations

$$V_t = V_B - V_a = \frac{kT}{e} \ln \frac{n_n}{n_p} = \frac{kT}{e} \ln \frac{p_p}{p_n}$$

(2-64)

where the carrier densities are no longer equilibrium values. We may solve Eq. (2-64) to obtain the values of minority-carrier concentration on either side of the junction:

$$n_p = \left[n_n \exp \left(-\frac{eV_B}{kT} \right) \right] \exp \frac{eV_a}{kT}$$

$$p_n = \left[p_p \exp \left(-\frac{eV_B}{kT} \right) \right] \exp \frac{eV_a}{kT}$$

(2-65)

The terms in the brackets may be identified with the equilibrium values of minority-carrier concentration, as may be verified by setting $V_a = 0$. Thus we obtain

$$n_p = n_{p_0} \exp \frac{eV_a}{kT}$$

$$p_n = p_{n_0} \exp \frac{eV_a}{kT}$$

(2-66)

and the *excess* values of minority-carrier concentration are

$$\hat{n}_p = n_p - n_{p_0} = n_{p_0} \left(\exp \frac{eV_a}{kT} - 1 \right)$$

$$\hat{p}_n = p_n - p_{n_0} = p_{n_0} \left(\exp \frac{eV_a}{kT} - 1 \right)$$

(2-67)

Equations (2-67) are extremely important in the theory of diodes and transistors, and are termed collectively the "law of the junction." They complete the necessary mathematical description required to discuss diodes and transistors. We should note one important derivation to be made from these equations. Refer for a moment to Eqs. (2-65). *These equations state that the minority-carrier concentration on one side of the junction is proportional to the majority-carrier concentration on the other side (i.e., the doping level on the other side).* Thus

$$\frac{n_p}{p_n} = \frac{n_n}{p_p} \approx \frac{N_D}{N_A}$$

(2-68)

In transistors, the emitter is heavily doped with respect to the base so that the injected minority-carrier concentration on the emitter side of the junction will be small with respect to that on the base side.

TABLE 2-1 Properties of Silicon

Atomic number..................	14
Atomic weight..................	28.06
Density (25°C)..................	2.33×10^3 kg/m³
Melting point..................	1420°C
Relative dielectric constant........	12

TABLE 2-2 Physical Constants

Magnitude of electron charge, e........	1.6×10^{-19} coulomb
Electron rest mass, m_e...............	9.11×10^{-31} kg
Planck's constant, h...............	6.62×10^{-34} joule-sec
Boltzmann's constant, k...............	1.38×10^{-23} joule/°K
Avogadro's number, N_0...............	6.02×10^{26} molecules/kg-mole
Permittivity of free space, ϵ_0...........	8.854×10^{-12} farad/m
Permeability of free space, μ_0..........	$4\pi \times 10^{-7}$ henry/m
Speed of light, C....................	3.00×10^8 m/sec

NOTE: $\dfrac{kT}{E} \approx 0.025$ volt at room temperature

$kT \approx 0.025$ ev at room temperature

1 ev = 1.6×10^{-19} joule

REFERENCES

Semiconductor Device Physics

1. Adler, R. B., A. C. Smith, and R. L. Longini: *Introduction to Semiconductor Physics*, Semiconductor Electronics Education Committee, vol. 1, John Wiley, 1964.
2. Jonscher, A. K.: *Principles of Semiconductor Device Operation*, John Wiley, 1960.
3. Lindmayer and Wrigley: *Fundamentals of Semiconductor Devices*, Van Nostrand, 1965.
4. Moll, J. L.: *Physics of Semiconductors*, McGraw-Hill, 1964.
5. Nussbaum, A.: *Semiconductor Device Physics*, Prentice-Hall, 1962.

PROBLEMS

2-1. Suppose an n-p-n transistor is constructed by uniformly doping three adjacent regions of intrinsic silicon with the following impurity levels:

$$
\begin{aligned}
&\text{Emitter:} &&N_D = 10^{19}/\text{cm}^3 \\
&\text{Base:} &&N_A = 10^{16}/\text{cm}^3 \\
&\text{Collector:} &&N_D = 10^{15}/\text{cm}^3
\end{aligned}
$$

Assuming all impurities are ionized, determine the equilibrium hole and electron concentrations in the regions at room temperature.

2-2. In the transistor above, assume that the lifetime for electrons in the base region is $\tau = 10^{-7}$ sec. Suppose an instantaneous burst of gamma radiation ionizes 10^9 silicon atoms/cm³. On the basis of recombination alone, determine the time required for the base region minority-carrier concentration to recover to within 10 percent of its equilibrium value. At this time, what will be the percentage of excess majority-carrier concentration?

2-3. The resistance of a material is given by

$$
R = \frac{\rho l}{A}
$$

where R = resistance, ohms
  ρ = resistivity, ohm-cm
  l = length, cm
  A = area, cm²

The resistivity of a semiconductor may be calculated from the equilibrium carrier concentrations by the relation

$$
\rho = \frac{1}{e(p\mu_p + n\mu_n)}
$$

where μ_p = hole mobility (480 cm²/volt-sec)
  μ_n = electron mobility (1,350 cm²/volt-sec)

Determine the resistivity of each of the three transistor regions of Prob. 2-1, and the resistance of a 0.1-cm-long slab with area $10^{-4}/$ cm² for each such region.

2-4. A p-n-p transistor is constructed by sandwiching a narrow n-type region between two p-type regions. The equilibrium minority-carrier concentration in the base region is $p_{n_0} = 10^5$ holes/cm³. Suppose the following voltages are applied to the transistor:

$V_{EB} = +0.25$ volt (forward direction)
$V_{CB} = -10.0$ volts (reverse direction)

Let $kT/e = 0.025$ volt. Determine the excess minority-carrier concentration in the base region at the edge of each junction (approximations are permitted).

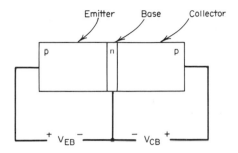

2-5. A two-lump model for the base region of a transistor is shown below. The symbols are the Linvill parameters for the base region. We have divided the base region into two parts so that $\Delta x_1 = \Delta x_2 = w/2$, where w is the width of the base region. Let

$$S_1 = S_2 = \frac{eAw}{2}$$

$$H_{C_1} = H_{C_2} = \frac{eAw}{2\tau}$$

$$H_D = \frac{eAD}{w}$$

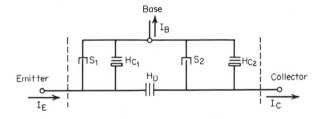

The transistor equations are

$$I_E = (H_D + H_{C_1})\hat{p}_E + S_1 \frac{d\hat{p}_E}{dt} - H_D \hat{p}_C$$

$$I_C = H_D \hat{p}_E - (H_D + H_{C_2})\hat{p}_C - S_2 \frac{d\hat{p}_C}{dt}$$

where $\hat{p}_E = p_{n_3}\left(\exp\frac{eV_{EB}}{kT}\right) - 1$

$$\hat{p}_C = p_{n_0}\left(\exp\frac{eV_{CB}}{kT}\right) - 1$$

The "dc common-base current gain" (α) of the transistor is defined as

$$\alpha = \frac{I_C}{I_E}\bigg|_{V_{CB}\,=\,0}$$

Assuming steady-state conditions ($d/dt = 0$), derive an expression for α in terms of geometrical and physical constants.

The "dc common-emitter current gain" (β) is defined as

$$\beta = \frac{I_C}{I_B}\bigg|_{V_{CB}\,=\,0}$$

For the current direction shown in the diagram,

$$I_E = I_C + I_B$$

Derive an expression for β in terms of α and in terms of geometrical and physical constants. What steps can be taken during fabrication to increase the gain?

THREE

Semiconductor Components

We continue our review of physical principles in this chapter with discussion of the operating mechanisms and characteristics of components employed in integrated circuits. The basic physics of the last chapter will be applied with appropriate boundary conditions to identify structures with useful device properties.

Diodes

Current-Voltage Characteristics. A diode is simply a p-n junction. An ideal diode acts as a short circuit when voltage is applied in the forward direction (positive terminal on the p side) and as an open circuit when voltage is applied in the reverse direction. Forward biasing for a diode is illustrated in Fig. 3-1a. A profile of the excess minority-carrier densities on either side of the junction is shown in Fig. 3-1b. We take the depletion region to extend from $x = -a$ to $x = b$. The width of the semiconductor on either side of the junction is assumed to be large with respect to the appropriate diffusion lengths.

Under this condition the development of Eq. (2-38) applies and we have

$$\hat{p}(x) = \hat{p}(b) \exp\left(-\frac{x-b}{L_p}\right) \qquad \text{on n side} \qquad (3\text{-}1)$$

$$\hat{n}(x) = \hat{n}(-a) \exp\frac{x+a}{L_n} \qquad \text{on p side} \qquad (3\text{-}2)$$

The respective minority-carrier currents on either side of the junction are

$$j_p = -eD_p\frac{d\hat{p}}{dx} = e\frac{D_p}{L_p}\hat{p}(b)\exp\left(-\frac{x-b}{L_p}\right) \qquad \text{on n side} \qquad (3\text{-}3)$$

$$j_n = eD_n\frac{d\hat{n}}{dx} = e\frac{D_n}{L_n}\hat{n}(-a)\exp\frac{x+a}{L_n} \qquad \text{on p side} \qquad (3\text{-}4)$$

We now make the assumption of negligible recombination within the depletion region. If this is true, continuity conditions require that

$$j_n(-a) = j_n(b) \qquad \text{electron current continuous} \qquad (3\text{-}5)$$

Thus the minority-carrier current on the p side of the junction is equal to the majority-carrier current on the n side of the junction. If the total current conducted through the diode is to remain continuous, then

$$j = j_n(x) + j_p(x)$$

(a)

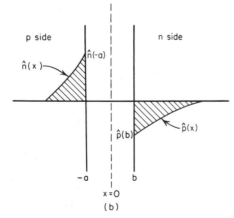

(b)

Fig. 3-1   *Diode with applied voltage. (a) Diode biased in forward direction. (b) Excess minority-carrier profile.*

is a constant for all x. Specifically, take $x = b$:

$$j = j_n(b) + j_p(b) = j_n(-a) + j_p(b)$$

$$= e\left[\frac{D_n}{L_n}\,\hat{n}(-a) + \frac{D_p}{L_p}\,\hat{p}(b)\right] \tag{3-6}$$

But from the law of the junction, Eqs. (2-66) and (2-67),

$$\hat{n}(-a) = n_{p0}\left(\exp\frac{eV_a}{kT} - 1\right) \approx \frac{n_i^2}{N_A}\left(\exp\frac{eV_a}{kT} - 1\right)$$

$$\hat{p}(b) = p_{n0}\left(\exp\frac{eV_a}{kT} - 1\right) \approx \frac{n_i^2}{N_D}\left(\exp\frac{eV_a}{kT} - 1\right)$$

so that

$$j = en_i^2\left(\frac{D_n}{L_n}\frac{1}{N_A} + \frac{D_p}{L_p}\frac{1}{N_D}\right)\left(\exp\frac{eV_a}{kT} - 1\right) \tag{3-7}$$

Using $I = Aj$, $L_n = (D_n\,\tau_n)^{1/2}$, and $L_p = (D_p\,\tau_p)^{1/2}$, we obtain the desired current-voltage characteristic:

$$I = I_0\left(\exp\frac{eV_a}{kT} - 1\right) \tag{3-8}$$

where

$$I_0 = eAn_i^2\left[\left(\frac{D_p}{\tau_p}\right)^{1/2}\frac{1}{N_D} + \left(\frac{D_n}{\tau_n}\right)^{1/2}\frac{1}{N_A}\right] \tag{3-9}$$

In quantity I_0 is called the diode reverse-saturation current (or leakage current). It is a small reverse current that flows when a large negative-bias voltage is applied (large with respect to $kT/e \approx 25$ mv). When a forward bias is applied, the current increases rapidly (exponentially) and has the appearance of a short circuit.

Thin-base Diodes. The current-voltage characteristic of a diode was derived in the last section under the assumption of wide base regions on either side of the p-n junction. We have seen that the excess minority-carrier concentration decays exponentially from the junction under this condition [Eqs. (3-1) and (3-2)]. It is often desirable to make the width of the semiconductor on one or both sides of the junction small with respect to a diffusion length. The profile of excess carriers is then linear, in general agreement with Eq. (2-41). If one carries through the previous argument for this case, the result is identical to that of Eq. (3-7), with the exception that the diffusion lengths on either side of the junction must be replaced by the actual base widths. If w_n is the actual width through which electrons diffuse on the p side and w_p is the corresponding width on the n side, Eq. (3-7) is changed to read

$$j = en_i^2 \left(\frac{D_n}{w_n} \frac{1}{N_A} + \frac{D_p}{w_p} \frac{1}{N_D} \right) \left(\exp \frac{eV_a}{kT} - 1 \right) \tag{3-10}$$

NOTE: The subscripts for D, L, and w refer to the minority-carrier type, and not to the material type.

Junction Capacitance.   The depletion region on either side of a p-n junction contains a net uncompensated charge whose magnitude depends upon the width of the depletion region, Eq. (2-53). The width of the depletion region depends, in turn, upon the magnitude of the applied voltage, Eqs. (2-61). We thus have a voltage-dependent mechanism for storing charge, or a capacitance effect.

By directing our attention to the indicated equations, it is observed that the depletion width, and hence the stored charge, is proportional to the square root of the net reverse potential $V_t = V_B - V_a$. Thus to increase the stored charge we must increase the applied bias V_a in the reverse direction. We define junction capacitance by the relation

$$C = -\frac{dQ}{dV_a} \tag{3-11}$$

From Eqs. (2-53) and (2-61) we have

$$Q = eAN_D\, d_n$$

$$= eAN_D \left[\frac{2\epsilon(V_B - V_a)}{e} \frac{N_A}{N_A N_D + N_D^2} \right]^{1/2} \tag{3-12}$$

and therefore

$$C = -\frac{dQ}{dV_a} = \left[\frac{A^2 \epsilon e N_D N_A}{2(V_B - V_a)(N_D + N_A)} \right]^{1/2} \tag{3-13}$$

Unlike the case of conventional capacitors, here junction capacitance is not constant with voltage, but instead decreases with increasing reverse bias. We may express Eq. (3-13) differently:

$$C = \frac{C_0}{1 - (V_a/V_B)^{1/2}} \tag{3-14}$$

where C_0 is the capacitance with zero applied bias. The reader may wonder about the applicability of Eqs. (3-13) and (3-14) for positive values of applied bias. In particular, it appears that an infinite value of capacitance would result from a positive applied voltage equal to the built-in voltage V_B. This is a fault in analysis resulting from the assumption that all of the applied voltage appears across the junction. With a positive applied bias, current will be drawn and an ohmic potential drop in the diode base region will result. That portion of the applied voltage appearing across the junction cannot, in fact, equal the built-in voltage before destructive changes in diode properties occur.

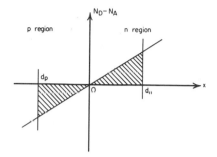

Fig. 3-2 Linear-graded impurity characteristic of diffused junction.

In our discussion of p-n junctions thus far, we have assumed a constant impurity doping on either side of the junction with an abrupt transition between two material types. In actual practice, the transition from p-type to n-type often takes place gradually in an almost linear fashion as shown in Fig. 3-2. Consideration of the capacitance of such a junction yields this modified equation (see Prob. 3-2):

$$C = \frac{C_0}{1 - (V_a/V_B)^{1/3}} \tag{3-15}$$

Breakdown Voltage. Under negative-bias voltage conditions, the potential appearing across the junction is the sum of the built-in and applied

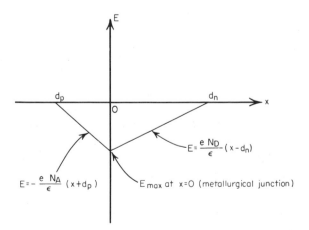

Fig. 3-3 Electric field in depletion region.

voltages. The bulk of the diode is essentially field-free. In accordance with Eqs. (2-56) and (2-57), the electric field in the depletion region varies linearly* with distance from the metallurgical junction as shown in Fig. 3-3. The maximum value of an electric field, occurring at the metallurgical junction, is

$$E_{max} = -\frac{eN_D\,d_n}{\epsilon} = -\frac{eN_A\,d_p}{\epsilon} \tag{3-16}$$

Substituting Eqs. (2-61) into (3-16), we obtain

$$E_{max} = -\frac{eN_D}{\epsilon}\left[\frac{2\epsilon(V_B - V_a)}{e}\frac{N_A}{N_A N_D + N_D^2}\right]^{1/2}$$

or

$$E_{max} = \left[\frac{2e(V_B - V_a)}{\epsilon}\frac{N_A N_D}{N_A + N_D}\right]^{1/2} \tag{3-17}$$

As the reverse bias is increased to large values, eventually a point will be reached where the material can no longer sustain the maximum electric field. The material "breaks down" and a large reverse current results. If we denote the characteristic field at which breakdown occurs as E_{br} and the corresponding applied voltage as V_{br}, we obtain from Eq. (3-17)

$$V_{br} = \frac{\epsilon}{2e}\frac{N_A + N_D}{N_A N_D}E_{br}^2 \tag{3-18}$$

In writing Eq. (3-18) we have assumed that $V_{br} \gg V_B$. Characteristic values for E_{br} are on the order of 0.3 to 0.5 million volts/cm.

Effects of Varying Impurity Concentration on Diode Parameters. We are now in a position to examine the effects of varying impurity concentration on important diode parameters. We consider the following parameters:

1. Ohmic loss in diode base (base resistivity)
2. Leakage current
3. Junction capacitance
4. Breakdown voltage

The first parameter, which we have not yet explicitly discussed, is the voltage drop across the diode bulk due to the finite resistance of the semiconductor and the current during forward conduction. It represents a departure from ideal diode characteristics and should be minimized. The voltage drop is proportional to the resistivity, Eq. (2-33).

To simplify the discussion, we assume that the p region is doped much

---

*When seeking a qualitative understanding of diode properties, we shall find it convenient to use results computed for the abrupt p-n junction because of the simplicity of the relationships involved. Equations (2-56) and (2-57) are not correct for the linear-graded junction, but the results obtained are useful for determining the process variables that affect the breakdown voltage.

more heavily than the n region ($N_A \gg N_D$). Under this condition, the following approximate expressions may be written for the parameters of interest:

1. Ohmic loss

$$\rho = \frac{1}{e\mu_n N_D} \tag{3-19}$$

2. Leakage current

$$I_0 = eAn_i^2 \left(\frac{D_p}{\tau_p} \frac{1}{N_D}\right)^{1/2} \tag{3-20}$$

3. Junction capacitance

$$C = \left[\frac{A^2 \epsilon e}{2(V_B - V_a)} N_D\right]^{1/2} \tag{3-21}$$

4. Breakdown voltage

$$V_{br} = \frac{\epsilon}{2e} E_{br}^2 \frac{1}{N_D} \tag{3-22}$$

In normal diode operation it is desirable to minimize the ohmic loss, leakage current, and junction capacitance and to maximize the breakdown voltage. It is apparent that we must be prepared to accept some compromise in adjusting the doping level, since the first two parameters can be optimized only at the expense of the last two.

Bipolar Transistors

General Discussion. The bipolar or conventional transistor is a three-terminal device which is capable of amplifying signals and acting as a

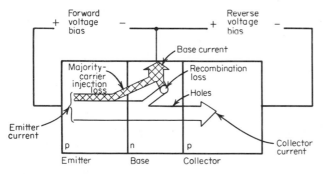

Fig. 3-4 Current flow in p-n-p transistor.

switch. The latter property is utilized in the construction of digital circuitry.

The mechanisms of current flow in a transistor are illustrated in Fig. 3-4. The transistor consists of three adjacent homogeneous regions of alternate doping, with a terminal contact attached to each region. A p-n junction exists at the interface between the emitter and base and between the base and collector. The width of the base is made small, though shown in exaggerated form in the figure.

We have discussed previously the current-flow process through a diode. As we have seen, the application of a forward voltage bias to a diode results in current injection across the p-n junction. The total current is the sum of two contributions: holes injected from the p side to the n side and electrons injected from the n side to the p side [see Eq. (3-6) and discussion leading to it]. This process, occurring across the emitter-base region, is illustrated in Fig. 3-4 for a p-n-p transistor. The electron current, injected from base to emitter, is shown crosshatched. It is desirable to minimize the electron component because only the hole component will diffuse across the base to the collector. The electron component, which is the majority-carrier component in the base, represents a loss to the total current transported. The fraction of the current injected across the emitter-base junction that consists of minority carriers (holes) is called the emitter efficiency (γ).

The base-collector junction is reverse-biased, and this condition constrains the minority-carrier concentration at the far side of the base to zero (see Prob. 2-4). A concentration gradient of minority carriers, therefore, exists across the base, and acting under the gradient the holes diffuse across the base to the base-collector junction. In the flow across the base, however, a fraction of the holes will recombine with electrons and will be lost to the current-flow process. The fraction of hole current surviving in the base is called the base transport factor (α'). The remaining hole current reaches the base-collector junction and is swept across the junction by the existing electric field, which is in such direction as to accelerate the holes into the collector. The hole current crossing into the collector is equal to the collector current.

As a result of the above discussion, two important points are evident:

1. The base current is solely majority-carrier current and is exactly equal to the current lost in transversing from emitter to collector.

2. The ratio of collector to emitter current is given by the product of the emitter efficiency and the base transport factor:

$$\frac{I_C}{I_E} = \alpha = \alpha'\gamma \tag{3-23}$$

Emitter Efficiency. The emitter efficiency is the ratio of minority current to the total current at the emitter-base junction:

$$\gamma \equiv \frac{I_{\min}}{I_{\min} + I_{\text{maj}}} = \frac{I_p}{I_p + I_n} = \frac{j_p}{j_p + j_n} \tag{3-24}$$

From Eqs. (3-3) and (3-4) we have

$$j_p = e \frac{D_p}{w} \hat{p}(0) \tag{3-25}$$

$$j_n = e \frac{D_n}{w} \hat{n}(0) \tag{3-26}$$

where w = base width, used in Eq. (3-25) because $w \ll L_p$ [see discussion associated with Eq. (3-10)]

$\hat{p}(0)$ = excess minority-carrier concentration at base side of junction

$\hat{n}(0)$ = excess minority-carrier concentration at emitter sides of junction

Also, from the law of the junction [Eq. (2-66)],

$$\hat{n}(0) = \frac{n_i^2}{N_A} \left(\exp \frac{eV_{EB}}{kT} - 1 \right)$$
$$\hat{p}(0) = \frac{n_i^2}{N_D} \left(\exp \frac{eV_{EB}}{kT} - 1 \right) \tag{3-27}$$

Substituting Eqs. (3-25) to (3-27) into Eq. (3-24), we obtain

$$\gamma = \frac{eD_p/wN_D}{eD_p/wN_D + eD_n/L_nN_A}$$

or

$$\gamma = \frac{1}{1 + D_nN_Dw/D_pN_AL_n} \tag{3-28}$$

To maximize the emitter efficiency, the doping concentration in the emitter is made much greater than that in the base ($N_A \gg N_D$). The emitter concentration is often indicated by the symbols n⁺ or p⁺ depending upon the transistor type. The plus sign indicates heavy doping. Quite often in integrated-circuit structures, the emitter width is also quite small and should be used instead of the diffusion length in Eq. (3-28).

A more general expression for the emitter efficiency, valid for both p-n-p and n-p-n transistors, is

$$\gamma = \frac{1}{1 + D_EN_Bw/D_BN_EL_E} \tag{3-29}$$

where D_E and D_B = diffusion constants for the minority carriers in the emitter and base regions, respectively

N_E and N_B = impurity concentrations in the emitter and base regions, respectively

w = base width

L_E = diffusion length of minority carriers in the emitter

Base Transport Factor. Neglecting the junction loss, the Linvill equations for transistor operation are (see Prob. 2-5)

$$I'_E = (H_D + H_{C_1})\hat{p}_E + S_1 \frac{d}{dt}\hat{p}_E - H_D\hat{p}_C$$

$$I_C = H_D\hat{p}_E - (H_D + H_{C_2})\hat{p}_C - S_2 \frac{d}{dt}\hat{p}_C$$

(3-30)

Here I'_E is the minority-carrier current entering the emitter side of the base region. Since we are interested in the steady-state solution, let $d/dt = 0$. Also $|\hat{p}_C| \approx |-p_{n_0}| \ll |\hat{p}_E|$ (see Prob. 2-4). Under these conditions, the base transport factor may be expressed as

$$\alpha' = \frac{I_C}{I'_E} = \frac{H_D}{H_D + H_{C_1}} = \frac{1}{1 + w^2/2L_p^2}$$

(3-31)

The total gain for the case of injection of emitter current (common-base current gain) is the product of expressions (3-28) and (3-31). The gain is a strong function of base width w and the ratio of the doping levels, N_D/N_A. For our purposes, we will assume that the emitter efficiency is unity, so that $\alpha = \alpha'$.*

Common-emitter Current Gain. In most applications of interest to our study, the transistor emitter will be grounded and the base terminal will be used to control the collector current. We are, therefore, interested in the common-emitter current gain (see Prob. 2-5):

$$\beta = \frac{I_C}{I_B} = \frac{\alpha}{1 - \alpha} = \frac{2L_p^2}{w^2}$$

(3-32)

Values of β range typically from 20 to 200. Thus, a small base input can control a large collector current.

Cutoff Frequency. We have thus far tacitly assumed a steady-state direct-current situation for the transistor. In many cases, it is of interest to amplify a periodic sinusoidal signal by applying the signal to the base terminal and making use of the output signal at the collector. From a qualitative standpoint, we know that it takes a finite time for carriers to diffuse across the base region. If the period of the sinusoidal signal becomes so small as to be comparable to this transport time, the polarity of the signal will change before all the carriers have been collected. We thus expect the gain to decrease as frequency increases (decreasing period of oscillation).

---

*In common integrated-circuit structures both α' and γ, as defined, can be made extremely close to unity. However, in such cases, other effects which have not been discussed become limiting constraints on these parameters. These effects include surface recombination and recombination within the depletion region.

Let us first consider the common-base current gain. We again refer to the Linvill equations. With a periodic sinusoidal signal, we may replace the time derivative d/dt with the complex frequency $j\omega$. The boundary conditions still require $\dot{p}_C \approx 0$. Under these conditions, we obtain from Eq. (3-30)

$$\alpha(\omega) = \frac{H_D}{H_D + H_{C_1} + j\omega S_1} \tag{3-33}$$

The cutoff frequency ω_α is defined as that frequency where the magnitude of $\alpha(\omega)$ has decreased to $\sqrt{2}/2$ of its dc value. At this frequency point, $\omega = \omega_\alpha$ and

$$\alpha(\omega_\alpha) = \frac{H_D}{H_D + H_{C_1} + j\omega_\alpha S_1} = \frac{\sqrt{2}}{2} \frac{H_D}{H_D + H_{C_1}} \tag{3-34}$$

Solving Eq. (3-34) for ω_α, we obtain

$$\omega_\alpha = \frac{H_D + H_{C_1}}{S_1} = \frac{(eAD_p/w) + (eAw/2\tau_p)}{eAw/2} \qquad \text{rad/sec} \tag{3-35}$$

In the usual case, $H_{C_1} \ll H_D$, so that

$$\omega_\alpha \approx \frac{2D_p}{w^2} \qquad \text{rad/sec} \tag{3-36}$$

and

$$f_\alpha = \frac{\omega_\alpha}{2\pi} = \frac{D_p}{\pi w^2} \qquad \text{Hz} \tag{3-37}$$

Let us compare this value of cutoff frequency with the transit time for carriers through the base region. The average velocity of the carriers is

$$v \approx \frac{j_p}{e p_{\text{av}}} \approx \frac{(eD_p/w)\hat{p}(0)}{e\hat{p}(0)/2} = \frac{D_p}{2w} \tag{3-38}$$

The transit time is therefore approximately

$$T \approx \frac{w}{v} = \frac{2w^2}{D_p} = \frac{1}{\omega_\alpha} \tag{3-39}$$

We next consider the cutoff frequency of the transistor when operated in the common-emitter configuration. Since $I_B = I_E - I_C$, we have from Eq. (3-30)

$$\beta(\omega) = \frac{I_C}{I_B} = \frac{H_D}{H_{C_1} + j\omega S_1} \tag{3-40}$$

The expression goes to $\sqrt{2}/2$ of its dc value when

$$\omega_\beta = \frac{H_{C_1}}{S_1} = \frac{1}{\tau_p} \tag{3-41}$$

Thus the common-emitter cutoff frequency is the reciprocal of the minority-carrier lifetime. Finally, note that

$$\frac{\omega_\alpha}{\omega_\beta} = \frac{2D_p/w^2}{1/\tau_p} = \frac{2D_p\tau_p}{w^2} = \frac{2L_p^2}{w^2} = \beta \tag{3-42}$$

The alpha cutoff frequency is greater than the beta cutoff frequency by a factor of β. The increased current gain (by a factor of β) of the common-emitter configuration entails a penalty in cutoff frequency (by a factor of β).

Graphical Representation of Transistor Characteristics. The important characteristics of transistor operation may be represented graphically, as shown in Fig. 3-5. This graph, called the collector characteristic, plots current as a function of collector voltage with base current as a parameter. A number of departures from our simplified theoretical discussion are evident. First, the collector current increases slightly with collector voltage, as shown by the small positive slope of the constant I_B lines. Thus the gain $\beta = I_C/I_B \approx 20$ is not precisely constant. Second, the curves "saturate" at low values of collector voltage and run into one another.

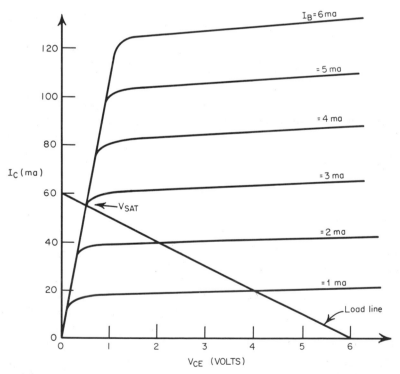

Fig. 3-5 *Graphic representation of transistor characteristics.*

A transistor switch is illustrated in Fig. 3-6. The operating characteristics of this switch are represented by the "load line" drawn into Fig. 3-5.
This line is the solution to the equation

$$V_{CE} = V_{CC} - I_C R_L = 6 \text{ volts} - (100 \text{ ohms}) I_C \tag{3-43}$$

The intersections of the constant I_B curves with the load line determine
the operating points. For example,

if $I_B = 1$ ma	then	$I_C = 20$ ma, $V_{CE} = 4$ volts
$I_B = 2$ ma		$I_C = 40$ ma, $V_{CE} = 2$ volts
$I_B = 3$ ma		$I_C = 55$ ma, $V_{CE} = 0.5$ volt
$I_B = 4$ ma		$I_C = 55$ ma, $V_{CE} = 0.5$ volt

Note the saturation of the characteristics at the higher values of base
current. As we increase the base current, V_{CE} decreases until the saturation
voltage is reached (denoted by V_{SAT} in Fig. 3-5). Further increases in base
current do not substantially change the collector current or voltage. The
transistor is said to be in "saturation." Note that the collector current can
be switched from 0 to 55 ma by a 3-ma change in base current.

Field-effect Transistors

General Discussion. Field-effect transistors are three-terminal devices which exhibit a variable resistance between two terminals, the source
and the drain. The resistance is controlled by a potential applied to a

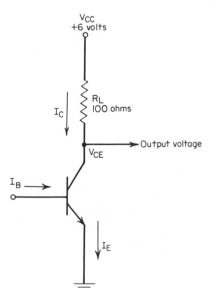

Fig. 3-6   Transistor   switch.

third terminal, the gate. The conducting region between the source and the drain is called the "channel." Current flow through a semiconductor resistor is composed mainly of majority carriers; the field-effect transistor thus differs profoundly in operating principle from the bipolar transistor just discussed. Since its operation involves only one carrier type, the field-effect device is often called the "unipolar" transistor.

Two principal construction schemes have been employed to realize practical devices. These schemes have resulted in the junction-gate field-effect transistor (FET) and the metal-oxide-semiconductor field-effect transistor (MOSFET) types. Of the two types, the characteristics of the MOSFET are more useful for digital-circuit applications, and this device has been employed to a significantly greater extent in integrated-circuit structures. We will therefore concentrate our attention on the MOS device.

The FET may be used as an amplifier or a switch, a resistor or a capacitor. When the FET is employed in integrated circuits, it is usually the only component which appears in the circuit. This procedure allows the construction parameters to be adjusted to optimize the device characteristics without concern for process compatibility with other integrated-circuit components. The construction simplicity of the MOSFET makes the device suitable for large circuit arrays involving subsystem-level functions (large-scale integration).

Basic Principles. The useful properties of MOSFETs are a consequence of the ability of an electric field to alter the distribution of free charge carriers in a semiconductor. Under certain circumstances, the conductivity type can be changed from n-type to p-type, or vice versa. Consider the capacitor structure of Fig. 3-7a, consisting of p-type silicon with subsequent overlayers of silicon dioxide (a dielectric) and a metal. The structure is similar to that of a parallel-plate capacitor if we imagine the silicon and metal layer to be the two plates.

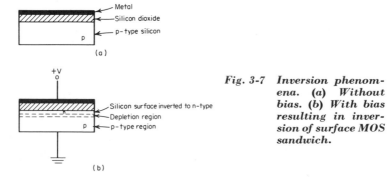

Fig. 3-7   *Inversion phenomena. (a) Without bias. (b) With bias resulting in inversion of surface MOS sandwich.*

The application of a voltage bias will result in charge being stored on the plates. Suppose a positive potential is applied to the upper metal plate as shown in Fig. 3-7b. The electric field established within the dielectric and terminated on negative charges within the semiconductor will attract electrons to the surface and repel holes (the former majority carrier) from the surface. At some magnitude of applied voltage, called the threshold voltage (V_{th}), a region near the surface of the semiconductor will become inverted, i.e., the conductivity type will change from p-type to n-type. An actual p-n junction will be formed between the n-type surface and the p-type base, and a depletion region will separate them.

It is possible to have an inverted surface channel existing prior to the application of voltage bias. This phenomenon is attributed to "surface states," and involves the interruption of the regular lattice pattern at the surface (dangling bonds) and the fields from extraneous ions existing within and around the dielectric. These encourage the enrichment of one carrier type over the other at the surface. In present-day structures, there is a tendency for positive ions to accumulate within the oxide, resulting in electron enrichment at the surface. With such a preexisting channel, a positive voltage bias will further extend the channel while a negative bias will diminish it. Such a device will therefore have a negative threshold voltage, commonly called the pinch-off voltage, V_{po}.

The above remarks apply equally well to a structure with an n-type base and p-type surface channel, except the sign of the voltages needs to be changed. A negative voltage will tend to form or further extend a surface channel and a positive-going voltage to reduce it.

Transistor device action is achieved by means of lateral current flow through the surface channel. An n-channel MOSFET is diagrammed in Fig. 3-8. The structure is similar to that of Fig. 3-7, except that electrical contact is made with both ends of the channel by means of two n-type contacting regions, the source and the drain. In terms of the coordinate system of the figure, the channel extends in the x direction into the semiconductor substrate to a depth d and has a y-directed length of l. If the channel is present, a potential applied between the drain and the source will cause current to flow between the two terminals. However, in the absence of the channel, conduction is not possible because one of the two p-n junctions which separate the source and the drain will be reverse-biased.

Current-Voltage Characteristics. The operating characteristics of the field-effect transistor may be examined by directing attention to the charge stored at the semiconductor surface. Assume that the transistor is biased with gate and drain voltages V_{GS} and V_{DS} and that the source is grounded, as shown in Fig. 3-8. The potential in the channel $V(y)$ is zero

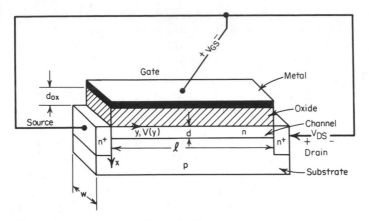

Fig. 3-8   n-channel MOS field-effect transistor.

at the source ($y = 0$) and equal to V_{DS} at the drain ($y = l$). The potential difference across the gate-to-substrate capacitor, $V_{GS} - V(y)$, is equal to the negative charge stored per unit area in the semiconductor, or

$$C_{ox}\left[V_{GS} - V(y) \right] = -q_{ss} - e\int_0^\infty (p - n + N_A)\, dx \qquad (3\text{-}44)$$

where
$$C_{ox} = \text{gate-to-substrate oxide capacitance per unit area}$$
$$= \frac{\epsilon_{ox}}{d_{ox}}, \text{ farads/cm}^2$$
$$q_{ss} = \text{surface-state charge per unit area, coulombs/cm}^2$$
$$e\int_0^\infty (p - n + N_A)\, dx = \text{charge stored in bulk per unit area, coulombs/cm}^2$$

The drain current may be found by integrating the electron drift current density across the channel area:

$$I_D = -I_y = -\int_{\substack{\text{channel}\\\text{area}}} j_y\, dA = -w\int_0^d j_y\, dx \qquad (3\text{-}45)$$

But $j_y = en\mu_n E_y = -en\mu_n\, [dV(y)/dy]$ [see Eq. (2-23)]; hence

$$I_D = e\mu_n w\left[\frac{dV(y)}{dy}\right]\int_0^d n\, dx \qquad (3\text{-}46)$$

In Eq. (3-46) we have assumed that the lateral electric field $E_y = -[dV(y)/dy]$ is not a function of x, a good approximation if the channel depth varies gradually. It may be shown that this condition requires the drain voltage to be small with respect to that magnitude of gate voltage in excess of the threshold value $(V_{DS} = V_{GS} - V_{th})$. From Eq. (3-44),

$$e \int_0^d n \, dx = C_{ox}\left[V_{GS} - V(y) \right]$$
$$+ q_{ss} - e \int_d^\infty n \, dx + e \int_0^\infty (p + N_A) \, dx \quad (3\text{-}47)$$

Defining

$$q_b = -e \int_d^\infty n \, dx + e \int_0^\infty (p + N_A) \, dx$$

and substituting Eq. (3-47) into (3-46), we obtain

$$I_D = w \left[\frac{dV(y)}{dy} \right] \mu_n \left\{ C_{ox}[V_{GS} - V(y)] + q_b + q_{ss} \right\} \quad (3\text{-}48)$$

when $V_{GS} - V(y) = V_{th}$, $I_D = 0$; therefore $-C_{ox}V_{th} = q_b + q_{ss}$, and Eq. (3-48) may be rewritten as

$$I_D = w\mu_n \frac{dV(y)}{dy} \left\{ C_{ox}[V_{GS} - V(y) - V_{th}] \right\} \quad (3\text{-}49)$$

Integrating Eq. (3-49) over the channel length, we obtain

$$I_D \int_0^l dy = w\mu_n C_{ox} \left\{ \int_0^{V_{DS}} [V_{GS} - V(y) - V_{th}] \, dV(y) \right\}$$

or

$$I_D = \frac{w}{l} \mu_n C_{ox} \left[(V_{GS} - V_{th})V_{DS} - \frac{V_{DS}^2}{2} \right] \quad (3\text{-}50)$$

The total gate-to-substrate capacitance C_g is

$$C_g = C_{ox}wl \quad (3\text{-}51)$$

which gives, finally,

$$I_D = \frac{\mu_n C_g}{l^2} \left[(V_{GS} - V_{th})V_{DS} - \frac{V_{DS}^2}{2} \right] \quad (3\text{-}52)$$

Equation (3-52) is the desired relationship between the drain current and the applied gate and drain voltage. It is valid for $V_{GS} > V_{th}$ and $V_{DS} \leq V_{GS} - V_{th}$. For $V_{GS} \leq V_{th}$, no drain current flows. For $V_{DS} > V_{GS} - V_{th}$, an approximate mathematical treatment of the device becomes quite difficult. The net gate-to-substrate voltage on the drain side of the channel, $V_{GS} - V_{DS}$, becomes less than the threshold voltage V_{th} and the channel becomes "pinched off" near the drain contact. Yet the higher value of drain voltage tends to encourage a greater value of drain current. What actually happens is that the channel assumes a grossly nonlinear shape and the current saturates to the value given by Eq. (3-52) for $V_{DS} = V_{GS} - V_{th}$:

$$I_D = I_{DS} = \frac{\mu_n C_g}{2l^2} (V_{GS} - V_{th})^2 \qquad (3\text{-}53)$$

Under conditions where Eq. (3-53) applies, the device is said to be in the saturation region of operation, and the drain current is called the saturation current.

An FET is characterized as either an enhancement-mode or a depletion-mode device, according to the sign of the threshold voltage. An enhancement-mode FET has a positive V_{th} and is nonconducting at zero gate bias. A depletion-mode FET has a negative V_{th} (called the pinch-off voltage, V_{po}) and conducts at zero gate bias.

Our discussion thus far has been directed to n-channel FETs. For p-channel devices, the signs of all voltage and current quantities must be inverted, and the hole mobility μ_p must be used in place of the electron mobility μ_n.

Device symbols, usual bias polarities, and sample drain characteristic curves for the four possible MOSFET types are shown in Fig. 3-9.

The MOSFET as a Circuit Element. The MOSFET is a high-input-impedance, voltage-controlled device, with characteristics strikingly similar to those of a pentode electron tube. The gate may be equated with the control grid, the source with the cathode, and the drain with the plate.

As an ac amplifier, the MOSFET is normally operated in the saturation region. A parameter of importance is the transconductance g_m, defined as the ratio of the incremental change in drain voltage to the incremental change in gate voltage. From Eq. (3-53)

$$g_m = \frac{\partial I_D}{\partial V_{GS}} = \frac{\mu_n C_g}{l^2} (V_{GS} - V_{th}) \qquad (3\text{-}54)$$

In the grounded-source configuration, the ac voltage gain is given by

$$A_v = -g_m R_L \qquad (3\text{-}55)$$

where R_L = effective load resistance in the drain circuit.

The frequency response of MOSFET is largely determined by the associated circuit impedances. The upper frequency at which the gain falls to $\sqrt{2}/2$ of its midband value (-3-db point) is given by

$$f_{3db} = \frac{1}{2\pi R_L C_O} \qquad (3\text{-}56)$$

where C_O is the effective output capacitance.

In digital applications, the MOSFET is employed as a switch. Enhancement-mode types are a natural choice for this purpose because the output of one enhancement-mode MOSFET is dc-compatible with the input requirements of another such device. These and other considerations are discussed in more detail in Chap. 8.

Resistors

Integrated-circuit resistors are simply long and narrow regions of semiconductor material which are electrically isolated from other components by a reverse-biased p-n junction. In the simplest case, the resistance is given by the relation

$$R = \frac{\rho l}{wd} \tag{3-57}$$

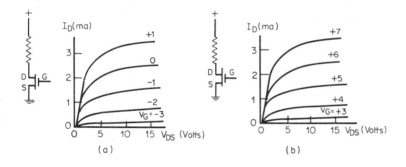

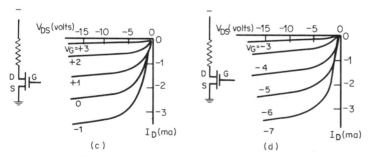

Fig. 3-9 *MOSFET types.* *(a) n-channel depletion.* *(b) n-channel enhancement.* *(c) p-channel depletion.* *(d) p-channel enhancement.*

where ρ = resistivity, ohm-cm
l = length of resistor, cm
w = width of resistor, cm
d = depth of resistor, cm

The resistivity may be determined from Eq. (2-33).

We shall soon see, however, that the impurity concentration, and hence the resistivity, is not constant, but varies with the depth into the material. In practice, it is common to use a "surface resistivity" that takes into account the impurity profile and diffusion depth.

Consider the resistor shown in Fig. 3-10, with resistivity $\rho(x)$ varying according to the depth x. We may divide the resistor into incremental slabs of depth, dx, each parallel to the other. The incremental conductance is

$$dG = d\left(\frac{1}{R}\right) = \frac{w}{\rho(x)l}\, dx \tag{3-58}$$

Summing all of the parallel conductance elements, we obtain

$$\frac{1}{R} = \left(\int_0^d \frac{1}{\rho(x)}\, dx\right)\frac{w}{l} \tag{3-59}$$

The resistance is thus determined by the length-to-width ratio according to the relation

$$R = \rho_s \frac{l}{w} \tag{3-60}$$

where $\rho_s \equiv 1/\int_0^d [1/\rho(x)]\, dx$ is the surface resistivity in ohms.

Capacitors

Capacitors in integrated circuits appear in two different forms. One technique utilizes the depletion-region capacitance of a reverse-biased

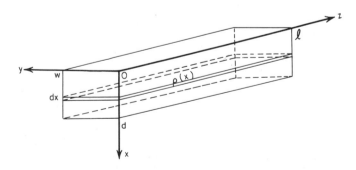

Fig. 3-10   Semiconductor region with variable resistivity.

p-n junction. The properties of such capacitors were developed with Eqs. (3-13) to (3-15).

A second technique employs the silicon dioxide passivation on the surface of the integrated circuit as a dielectric. Beneath the oxide is a low-resistivity diffused-semiconductor region which acts as one of the conductor plates. Above the oxide is a deposited metallic conductor which serves as the other plate. The capacitance is given by the familiar formula for a parallel-plate capacitor:

$$C = \frac{A\epsilon}{t} \tag{3-61}$$

where t is the oxide thickness.

REFERENCES

Transistors and Diodes

1. Gray, P. E., D. Dewitt, D. R. Boothroyd, and J. F. Gibbons: *Physical Electronics and Circuit Models of Transistors*, Semiconductor Electronics Education Committee, vol. 2, John Wiley, 1964.
2. Jonscher, A. K.: *Principles of Semiconductor Device Operation*, John Wiley, 1960.
3. Linvill, J. G., and J. F. Gibbons: *Transistors and Active Circuits*, McGraw-Hill, 1961.

Field-effect Transistors

4. Research Triangle Institute: *Integrated Silicon Device Technology, Volume VI. Unipolar Transistors*, Technical Documentary Report ASD-TDR-63-316, prepared for Air Force Systems Command, March, 1965.
5. Sah, C. T.: "Characteristics of the Metal-Oxide-Semiconductor Transistors," *IEEE Trans. Electron. Devices*, vol. ED-11, pp. 324–345, July, 1964.

PROBLEMS

3-1. The depletion region may be considered as the dielectric of a parallel capacitor whose thickness $d = d_n + d_p$. Using this approach and Eq. (3-61), calculate the junction capacitance and show that the result is identical to that of Eq. (3-13).

3-2. Diffused junctions in integrated circuits may be approximated by the linear grading (see Fig. 2-2)

$$N_D - N_A = ax$$

Under this condition we have $d_n = d_p = l$. The charge density is $\rho = e(N_D - N_A) = eax$.

a. Using Poisson's equation, Eq. (2-30), show that the electric field in the depletion region is given by

$$E = \frac{ea}{2\epsilon}(x^2 - l^2)$$

   b. Integrate the result of part a over the depletion region to obtain the following expression for the potential across the junction:

$$V_t = \frac{2}{3}\frac{l^3 ea}{\epsilon}$$

and from this show that the width of the depletion region is given by

$$d = 2l = \left(\frac{12V_t\epsilon}{ea}\right)^{1/3}$$

   c. Use Eq. (3-61) to determine the depletion-layer capacitance and show that it may be expressed as

$$C = \frac{C_0}{(1 - V_a/V_B)^{1/3}}$$

   where $C_0 = (A^{3/2}ea/12V_B)^{1/3}$

$$V_B - V_a = V_t$$

3-3. Determine f_T, the frequency at which the magnitude of common-emitter gain is unity, in terms of transistor physical parameters. Making suitable assumptions, show that it is equal to f_α.

3-4. An n-channel enhancement-mode MOSFET has the following parameters:

$$C_g = 4 \text{ pf}$$
$$l = 4 \times 10^{-3} \text{ cm}$$
$$V_{th} = 2 \text{ volts}$$

With $V_{DS} = 6$ volts, determine the drain current at $V_{GS} = 3$, 6, and 9 volts. Determine the transconductance in each case.

FOUR

Basic Construction Techniques

At this point in our study we depart from our consideration of device physics and proceed to develop the ideas underlying the integrated construction of a number of semiconductor components within a homogeneous chip of semiconductor material. The importance of doping and geometry control in the formation of useful devices has been discussed in previous sections. The techniques of photolithographic masking and of impurity diffusion from the semiconductor surface make possible the rigid controls required.

Our study of basic construction techniques will begin with consideration of the mechanism of impurity diffusion. Equations will be derived, subject to usual boundary conditions, which will relate the impurity concentration, as a function of distance into the material, to time and temperature processing variables. We will then discuss how a p-n junction is formed by a sequence of two diffusions.

Following our study of diffusion technology, we will consider a number of ways in which integrated-circuit structures are currently constructed.

Present-day techniques involve the additional formation of single-crystal semiconductors from the vapor phase, and this will lead us to a discussion of epitaxial growth processes. Finally, we will discuss the photoetching procedures involved in the oxide-masking steps.

Impurity Diffusion

The diffusion of impurities into semiconductor material involves the application of a given concentration of impurity atoms at the surface of the semiconductor in a high-temperature environment. The mechanism that governs the diffusion of mobile charge carriers from regions of high concentration to regions of low concentration is equally applicable to the diffusion of impurity atoms. We recall that the flux of particles (particle current) is proportional to the concentration gradient:

$$f = -D \frac{\partial N}{\partial x} \tag{4-1}$$

where D = diffusion coefficient, cm²/sec
 f = particle current, particles/cm²-sec
 N = particle concentration, particles/cm³
Continuity conditions require that [compare to Eq. (2-27) without recombination term]

$$\frac{\partial N}{dt} = -\frac{\partial f}{\partial x} \tag{4-2}$$

Combining Eqs. (4-1) and (4-2), we obtain the equation governing diffusion, called Fick's second law:

$$\frac{\partial N}{\partial t} = D \frac{\partial^2 N}{\partial x^2} \tag{4-3}$$

Our study of diffusion processes will involve the solution of Eq. (4-3) under various boundary conditions.

Use of Laplace Transforms. The solution of Eq. (4-3) can be facilitated by employing Laplace transform techniques. The Laplace transform of a variable is an operator defined by the operation

$$\mathcal{L}\{N(x,t)\} \equiv \bar{N}(x,s) \equiv \int_0^\infty \exp(-st)\, N(x,t)\, dt \tag{4-4}$$

The use of Laplace transforms in partial differential equations is based on five basic theorems, listed as follows:

 Theorem 1 $\mathcal{L}\{N_1 + N_2\} = \bar{N}_1 + \bar{N}_2$

 Theorem 2 $\mathcal{L}\left\{\dfrac{\partial N}{\partial t}\right\} = s\bar{N} - N(x,0)$

 Theorem 3 $\mathcal{L}\left\{\dfrac{\partial^n N}{\partial x^n}\right\} = \dfrac{\partial^n \bar{N}}{\partial x^n}$

Theorem 4 $\quad \mathcal{L}\left\{\int_0^t N(x,\tau)d\tau\right\} = \frac{1}{s}\bar{N}(x,s)$

Theorem 5 $\quad \mathcal{L}\{N(x,kT)\} = \frac{1}{k}\bar{N}\left(x,\frac{s}{k}\right) \qquad k > 0$

Diffusion of Impurities from a Constant Source. Consider the process conditions shown in Fig. 4-1. The surface of the semiconductor is maintained at a constant impurity-atom concentration, perhaps in gaseous form. The width of the semiconductor is large compared with the distances through which the impurities will diffuse, and may for all practical purposes be considered infinite. The boundary conditions are

$$N(\infty,t) = 0 \tag{4-5}$$

$$N(0,t) = N_0 \tag{4-6}$$

Using Ths. 2 and 3, the Laplace transform of Eq. (4-3) is

$$s\bar{N}(x,s) - N(x,0) = D\frac{d^2\bar{N}}{dx^2} \tag{4-7}$$

If the diffusion is assumed to begin at $t = 0$, the initial condition is

$$N(x,0) = 0 \tag{4-8}$$

Thus the transformed diffusion equation is

$$s\bar{N} = D\frac{d^2\bar{N}}{dx^2} \tag{4-9}$$

In obtaining the solution of $\bar{N}$ as a function of x, we may treat the Laplace variable s as a constant. Equation (4-9) is a common differential form whose solution is

$$A \exp x\sqrt{\frac{s}{D}} + B \exp\left(-x\sqrt{\frac{s}{D}}\right) \tag{4-10}$$

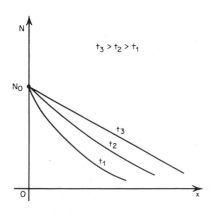

Fig. 4-1 *Diffusion from a constant source.*

where A and B are constants to be determined from the boundary conditions. We must, however, transform the boundary conditions (4-5) and (4-6) to obtain*

$$\bar{N}(\infty,s) = 0 \tag{4-11}$$

$$\bar{N}(0,s) = \frac{N_0}{s} \tag{4-12}$$

Condition (4-11) requires that $A = 0$ and condition (4-12) that $B = N_0/s$. Thus our solution is

$$\bar{N}(x,s) = \frac{N_0}{s} \exp\left(-x\sqrt{\frac{s}{D}}\right) \tag{4-13}$$

The inverse transform of Eq. (4-13) is†

$$N(x,t) = N_0 \operatorname{erfc} \frac{x}{2\sqrt{Dt}} \tag{4-14}$$

where

erfc y = complementary error function of y

$$= 1 - \operatorname{erf} y$$

$$= \frac{2}{\sqrt{\pi}} \int_y^\infty e^{-Z^2}\, dZ \tag{4-15}$$

Graphs of erf y and erfc y are drawn in Fig. 4-2.

As shown in Fig. 4-1, the concentration at a given value of x into the semiconductor increases with time, the profile being the complementary error function. For small values of their arguments, erf y and erfc y are almost linear, as demonstrated in the following table. The linear graded approximation of Prob. 3-2 is thus reasonable.

y	erf y	erfc y
0.0	0.00	1.00
0.1	0.11	0.89
0.2	0.22	0.78
0.3	0.33	0.67
0.4	0.43	0.57
0.5	0.52	0.48
0.6	0.60	0.40
0.7	0.68	0.32
0.8	0.74	0.26
0.9	0.80	0.20
1.0	0.84	0.16

*Note that the Laplace transform of a constant C is C/s from Eq. (4-4).

†See for example: The Chemical Rubber Co., *Handbook of Chemistry and Physics*, 48th Edition, 1967, Laplace transform pair 83, p. A-236.

Junction Formation. In the construction of integrated circuits, junctions are formed by a number of successive diffusions. A possible structure for an integrated-circuit transistor is illustrated in Fig. 4-3. The original starting material is uniformly doped p-type with $N_{A_1} = 10^{14}/cm^3$. The first diffusion converts a region of the material to n-type, as shown by the N_{D_1} curve. The intersection of the N_{D_1} curve and the N_{A_1} line is the point of zero net doping, or the collector-substrate junction. This junction, normally back-biased, isolates the transistor from the other components of the circuit. Subsequent diffusions (N_{A_2} and N_{D_2}) form the base and

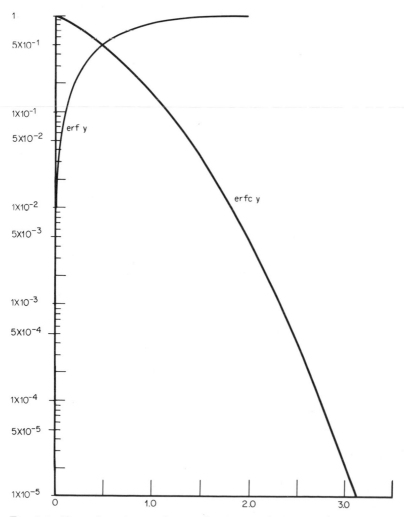

Fig. 4-2 Error function and complement as function of argument.

emitter regions. Note that each subsequent diffusion (at a high temperature) allows the resumption of previous diffusions. Thus the collector-substrate junction first formed by the N_{D_1} and N_{A_1} lines is moved to the right as subsequent diffusions are carried out. The formation of junctions requires careful schedules of initial concentration, temperature, and diffusion time.

Temperature Dependence of Diffusion Coefficient. The diffusion of impurities, essentially zero at room temperature, is accelerated by high temperatures. The temperature dependence of diffusion rates arises from the sensitivity of the diffusion coefficient to temperature, given approximately by

$$D = D_0 \exp\left(-\frac{\Delta H}{kT}\right) \tag{4-16}$$

where D_0 = constant depending on dopant and material
$\quad\quad\quad T$ = temperature, °K
$\quad\quad\quad \Delta H$ = activation energy depending upon dopant and material

Typical diffusion temperatures are in the range between 1000 and 1300°C and the desired diffusion profiles are attained in a period of time ranging from several minutes to a day. With impurities commonly used, a 10 percent change in temperature can cause an order-of-magnitude change in the diffusion constant.

Outdiffusion. Often, in the construction of integrated circuits, additional silicon is grown upon the original starting material by a process known as epitaxy. The additional material assumes the same crystalline orientation as the starting material, and the resulting whole can be considered a single structure. If the starting material is doped, impurities will

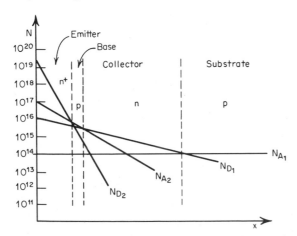

Fig. 4-3   Integrated-circuit-transistor junction formation by successive diffusions.

diffuse into the epitaxial film. Solution of the diffusion equation in a manner similar to the previous calculation (subject to the new boundary conditions) yields

$$N(x) = \frac{N_0}{2} \text{ erfc } \frac{x}{2\sqrt{Dt}} \qquad (4\text{-}17)$$

The diffusion profile is illustrated in Fig. 4-4.

Predeposit Diffusion. A technique often used in the construction of integrated circuits involves the deposition of a given number of impurity atoms per square centimeter upon the surface of the material to be doped. Impurities subsequently diffuse into the heated material, depleting the surface concentration. Since a constant concentration is not maintained at the surface during the diffusion, the surface concentration decreases with time, as illustrated in Fig. 4-5.

The diffusion profile is found from the diffusion equation to be

$$N(x,t) = \frac{M}{\sqrt{\pi Dt}} \exp\left(\frac{x^2}{4Dt}\right) \qquad (4\text{-}18)$$

where M = number of predeposited impurity atoms/cm².

Integrated-circuit Structures

Prevalent integrated-circuit structures are fabricated by one of five different techniques, listed as follows in historical order of introduction into the technology:

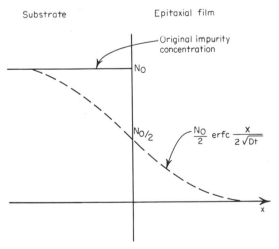

Fig. 4-4   *Outdiffusion into epitaxial film.*

1. Triple-diffused (diffused collector)
2. Triple-diffused (diffused isolation)
3. Epitaxial
4. Epitaxial with buried layer
5. Dielectrically isolated

We now discuss qualitatively the essential construction steps of these fabrication types, and the basic advantages and disadvantages of each type.

Triple-diffused (Diffused Collector). The process sequence for constructing a diffused-collector structure is illustrated in Fig. 4-6. Since all components are derived essentially from the basic bipolar structure, the formation of a transistor is used for illustrative purposes. Bear in mind that diodes are formed by using either of the transistor junctions, and resistors are usually long and narrow segments formed by the base diffusion. The construction of MOSFET circuits is a simple but separate problem which is discussed in the next chapter. The diffusion depths are exaggerated in the figure; they are usually less than a mil in a 5- to 10-mil-thick silicon wafer.

In Fig. 4-6a the original p-type starting material is shown. The wafer is oxidized to provide an SiO_2 diffusion barrier around the entire wafer (the diffusion coefficient of common impurities in SiO_2 is very much smaller than in Si). A photoresist material is next applied to the wafer, and certain areas are illuminated and "fixed." The unfixed photoresist is then dissolved in a photoresist solvent. The next step involves a hydrofluoric acid etch to remove the SiO_2 from the unfixed areas, for example over the desired collector region. The wafer is now ready for the first diffusion, as shown in Fig. 4-6b. Note that the diffused region spreads laterally (horizontally) under the oxide-protected region, as well as vertically. This phenomenon imposes a limit on obtainable surface-geometry miniaturization and tolerances.

After (or during) this n-type collector diffusion the wafer is reoxidized,

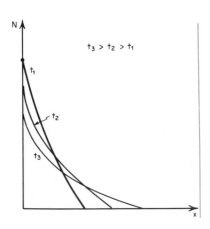

Fig. 4-5   *Predeposit diffusion profiles.*

and the above process is repeated to delineate the base regions. The second diffusion takes place as shown in Fig. 4-6c. The oxidation and photoetching procedure is repeated again in preparation for the emitter diffusion, which is shown in Fig. 4-6d. The high-concentration emitter diffusion is also allowed into the collector to form low-resistivity contact areas. Thus, four regions are formed: emitter, base, collector, and substrate. The substrate region is required to provide isolation of the components from one another. Any current flow from one component to the next must proceed through an n-p and p-n diode — one diode will always be back-biased.

The final step involves contacting to the components and interconnecting them in the desired circuit configuration by means of a thin-film aluminum pattern on the surface.

The diffused collector structure is the simplest and was, in the early

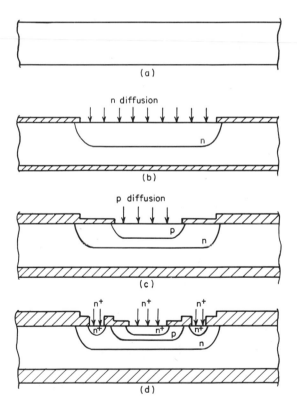

Fig. 4-6 Sequence for triple-diffused structure (diffused collector). (a) Original starting material. (b) Collector diffusion. (c) Base diffusion. (d) Emitter diffusion.

phases of the technology, the most economical of the integrated-circuit structures. The process steps are minimized and the required diffusion times are short. There are, however, certain disadvantages to the structure. The collector impurity concentration, formed by a diffusion, is graded. The higher concentration is at the base end of the collector and the lower concentration at the substrate end. This is just opposite to the desired result. High concentration at the base side lowers the base-collector breakdown voltage and increases the base-collector depletion capacitance. To obtain acceptable values for these two parameters, the overall concentration must be made low. This results in a high-resistivity collector with an undesirable ohmic loss (high V_{SAT}) and a limitation in frequency performance due to a high RC time constant at the collector. Better characteristics could be obtained if the collector could be uniformly doped with a carefully selected concentration. This is possible in the diffused-isolation structure.

Triple-diffused (Diffused Isolation). The first two steps in the diffused-isolation structure formation are illustrated in Fig. 4-7. Here the original starting material is n-type with a doping concentration optimized for the collector. For the first diffusion, the oxide is removed from all areas (top and bottom) *except* where components are to be formed. The first diffusion forms the p-type isolation region. Subsequent diffusions are as in Fig. 4-6.

The advantage of this technique is an optimized collector concentration. There are three disadvantages: first, the p diffusion must be made with high surface concentration for a long period of time with a thin (3 to 4 mils) and fragile wafer to permit diffusion throughout the wafer; second, the high

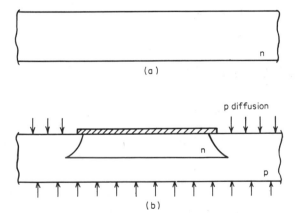

Fig. 4-7 *Collector formation in diffused-isolation structure. (a) Original starting material. (b) Isolation diffusion.*

surface concentration results in high collector-substrate capacitance which limits frequency performance; and third, precise control of collector depth is difficult because of the extent of diffusion required. To circumvent many of these difficulties, the epitaxial process may be employed.

Epitaxial Structure. The collector formation in the epitaxial structure is illustrated in Fig. 4-8. The original starting material is p-type, and will serve both as structural support and as an isolation region. The first step, Fig. 4-8*b*, involves the growth of an epitaxial silicon layer with uniform n-type doping suitable for the collector. Following this, an isolation diffusion converts regions of the epitaxial layer to p-type, leaving n-type collector and other component regions. Subsequent steps are similar to those previously discussed.

Epitaxial construction permits the advantages of the isolation-diffusion technique without requiring long diffusion times, high surface concentrations, and thin wafers. It does, however, add costly processing steps.

One problem remains for our attention. We have seen that a low collector concentration optimizes the collector-base breakdown voltage and depletion capacitance but gives high collector resistivity, and hence ohmic loss and limited frequency capabilities. If the collector concentration is

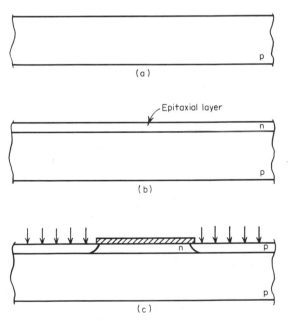

Fig. 4-8   Collector formation in epitaxial structure. (a) Original starting material. (b) Epitaxial-layer growth. (c) Isolation diffusion.

increased, we improve the latter at the expense of the former. Can we avoid this direct tradeoff problem? The answer is found in the epitaxial structure with a buried layer.

Epitaxial Structure with Buried Layer. The initial process steps for this structure are illustrated in Fig. 4-9. The procedures are identical to the previous structure with the exception of the deposition of an n+ predeposit over the collector area prior to epitaxial growth. During the epitaxial growth, the predeposit diffuses both into the epitaxial layer and substrate (Fig. 4-9c). Isolation diffusion and subsequent process steps are then performed as before.

The buried layer gives a low-resistivity region beneath the collector to minimize ohmic loss and increase frequency capabilities. The collector region itself can be made higher in resistivity to optimize collector-base breakdown voltage and depletion capacitance. This procedure, of course, adds cost to the overall integrated-circuit process.

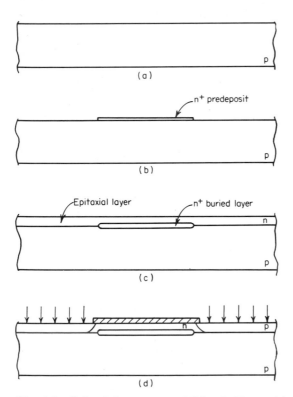

Fig. 4-9   Epitaxial structure with buried layer. (a) Original starting material. (b) n+ predeposit. (c) Epitaxial growth and buried-layer formation. (d) Isolation diffusion.

Dielectrically Isolated Structures. The use of the epitaxial structure with a buried layer provides good static characteristics and excellent frequency capabilities. The ultimate limitation in frequency capabilities is associated with the collector-to-substrate depletion capacitance which is invariably present with p-n junction isolation. The final process to be discussed, and also the most recent chronologically, replaces p-n junction isolation between components by a thick dielectric material, with a considerable reduction in capacitive coupling.

The initial process steps for a dielectrically isolated structure are diagramed in Fig. 4-10. The original starting material is a relatively thin, uniformly doped n-type wafer. Pits are acid-etched into the wafer and subsequently oxidized, as shown in Fig. 4-10b. In the next step, a low-

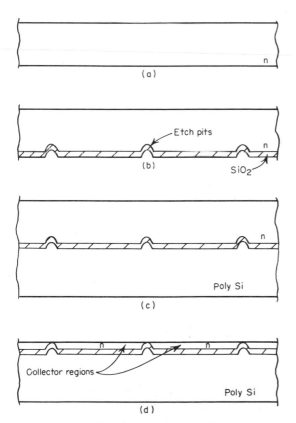

Fig. 4-10 Initial steps in dielectrically isolated structure. (a) Original starting material. (b) Formation of etch pits and subsequent oxidation. (c) Growth of poly-Si substrate. (d) Top surface lapped down to SiO_2.

quality polycrystalline layer is grown underneath the etched pits to give structural integrity to the finished product. The layer is made relatively thick, as shown in Fig. 4-10c. The upper surface is then lapped off to the SiO_2 layer, leaving isolated collector "islands" for subsequent processing. A buried layer may be incorporated as an additional feature. This type of construction can be quite expensive, but improvements in technology can be expected to lower production costs substantially.

Epitaxial Growth

Epitaxy is the deposition of silicon atoms onto a seed crystal (substrate) in such a manner that the added silicon assumes the same single-crystal organization and orientation as the seed crystal. Epitaxial growth may be accomplished by the hydrogen reduction of a silicon halide ($SiCl_4$ or $SiBr_4$) or the pyrolysis of silane (SiH_4) gas. In the former process, the seed crystal is maintained at an elevated temperature ($\approx 1200°C$) in a quartz reaction chamber by radio-frequency induction heating. A mixture of hydrogen and halide gas is made to pass over the heated substrate to obtain the reaction

$$SiX_4 + 2H_2 \xrightarrow{1200°C} Si + 4HX \tag{4-19}$$

where X represents a halide, usually Cl or Br. The silicon reaction product is deposited on the substrate, while the gaseous by-product passes through the chamber and is exhausted.

In the second process, silane is pyrolytically decomposed in contact with the heated substrate according to the reaction

$$SiH_4 \longrightarrow Si + 2H_2 \tag{4-20}$$

To achieve the desired doping levels, appropriate dopant gases may be added to the other reactants.

The pyrolytic decomposition of silane can be performed at a somewhat lower temperature ($\approx 1100°C$) than the hydrogen-reduction process. The lower reaction temperature is an advantage when we want to minimize the simultaneous diffusion of impurities during the reaction.

Photoetching Procedures

The use of oxide masking is made possible by highly precise photographic procedures. In the early days of integrated-circuit fabrication, poor tolerance in the film masks was a major cause of poor manufacturing yields.

The first step in photoetching is the construction of large-scale drawings which distinguish by opaqueness and transparency the oxide windows to be opened during the various diffusion steps. These drawings are photographed and reduced. It is well to note that as many as 1,000 complete circuits may be made on a single wafer. Thus the film must have the drawing repeated many times over on its surface. An error in alignment or registration of 5 microns for any one of the many circuits can cause serious problems.

In the next step, the wafer is treated with a photoresist, an organic emulsion which is soluble in a number of solvents, but which is polymerized, and hence made insoluble, by ultraviolet light. The photographic mask is placed over the treated wafer and aligned under a microscope. The wafer is then exposed to ultraviolet light, with the result that certain regions of the photoresist are fixed. The unexposed photoresist is then dissolved with a developer.

The oxide under the unexposed photoresist regions may now be removed with a hydrofluoric acid etch. Prior to diffusion, the polymerized photoresist is removed by a liquid abrasive or stripper.

REFERENCES

1. Research Triangle Institute: *Integrated Silicon Device Technology,* vol. IV: *Diffusion,* February, 1964; vol. IX: *Epitaxy,* August, 1965; vol. III: *Photoengraving,* January, 1964, Technical Documentary Report ASD-TDR-63-316, prepared for Air Force Systems Command.
2. Eastman Kodak Company: *Techniques of Microphotography,* Industrial Data Book P-52, 1963, and *Kodak Photo-Sensitive Resists for Industry,* Industrial Data Book P-7, 1962.

PROBLEMS

4-1. A sample of p-type silicon, originally uniformly doped with boron to a concentration of $10^{16}/cm^3$, is diffused with phosphorus at a constant surface concentration of $10^{18}/cm^3$.

    a. How long a diffusion at 1250°C will be required to produce a junction at a depth of 5×10^{-4} cm? ($D = 7.5 \times 10^{-12}$ cm²/sec at 1250°C for phosphorus.)

    b. Several years ago it was proposed to make a p-n-p structure by simultaneous diffusion of indium ($D = 2.8 \times 10^{-12}$ cm²/sec at 1250°C) with the phosphorus. What constant surface concentration would be needed to produce the first junction at a depth of 2.5×10^{-4} cm, assuming the diffusion time found in part *a?*

FIVE

Integrated-circuit Component and Intercomponent Characteristics

In the previous discussions, the first-order theory basic to the operation and construction of integrated-circuit components has been discussed. Although this theory is helpful for gaining an insight into the major factors that determine component performance, the realization of practical integrated-circuit structures involves additional complexities that are often handled by trial-and-error methods. The fabrication of integrated circuits makes use of many proprietary procedures, and the whole has often been likened to an art rather than to a science.

In the discussion that follows, an attempt will be made to summarize the present achievements and limitations of the technology with respect to important device parameters. The reader should bear in mind that the simultaneous processing of integrated circuits requires tradeoffs among the characteristics of the various components and their several parameters. In compromising among these parameters, and in the emphasis of proprietary-process advantages, wide differences appear in the product lines of manufacturers.

Integrated circuits employing bipolar transistors are treated separately

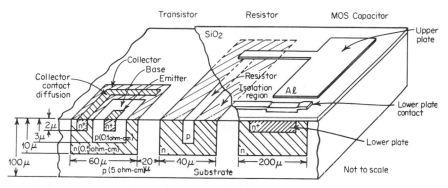

Fig. 5-1   Section of silicon integrated circuit using p-n junction.

from those composed of MOSFETs. At the present time, there is no great tendency to mix the two technologies.

Following the discussion of component characteristics and a few points concerning custom design, we briefly consider the component interaction made inevitable by close proximity in the substrate.

Characteristics of
Integrated-circuit Components

Bipolar Transistors. The bipolar transistor is perhaps the most perfected and highest-performance component in integrated circuits. Base-widths on the order of a fraction of a micron are typical. Development has reached the stage where bulk properties are no longer determinant in the characteristics of integrated circuits. Gain and frequency response are presently limited by second-order effects such as surface and depletion-region recombination, and RC time constants associated with the ohmic bulk and junction isolation capacitance.

A small section of an integrated circuit illustrating a typical geometrical arrangement of a transistor, resistor, and an MOS capacitor is shown in Fig. 5-1. The diffusion depths and spacings shown are fairly representative of prevalent integrated-circuit structures. The resistivity given for the base region should be considered as an average value, since the impurity concentration is, of course, not constant in the diffused structure.

The characteristics of integrated-circuit transistors do not differ significantly from their state-of-the-art conventional-circuit counterparts. One dissimilarity, except in structures using a buried layer, is a higher-than-normal collector ohmic drop arising from the resistance of the lateral path to the surface contacts (in conventional transistors, contact is made directly under the collector). Representative values for important transistor characteristics are listed in Table 5-1.

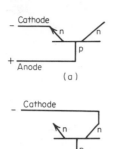

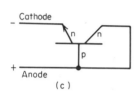

Fig. 5-2 Common diode connections. (a) Emitter-base diode. (b) Collector-base diode. (c) Diode-connected transistor configuration.

Diodes. Diodes, in integrated circuits, are formed from one of the two transistor junctions or some connected combination of the two. The three most prevalent connections are shown in Fig. 5-2. The emitter-base diode connection is shown in Fig. 5-2*a*. This diode has a low forward-voltage drop when conducting (≈ 0.8 volt at 2 ma) but suffers from a correspondingly low breakdown voltage (≈ 3 to 5 volts) due to the high emitter doping [see Eq. (3-22)]. Since a number of emitters may be diffused into a single base structure, the configuration is especially useful for arrays of common anode diodes.

An alternate diode configuration is the collector-base diode of Fig. 5-2*b*. The breakdown voltage of this diode is considerably greater (≈ 10 volts) than that of the previous case. The forward voltage, however, is also greater (≈ 1.0 volt at 2 ma). The structure lends itself well to common cathode configurations, since a number of base regions can easily be diffused into a single collector region.

The diode-connected transistor configuration (Fig. 5-2*c*) provides the lowest forward-voltage drop of any of the three connections (≈ 0.7 volt at 2 ma). The low voltage drop results from the transistor action of the connection. The breakdown voltage is comparable to that of the emitter-base diode. A disadvantage of the connection over the emitter-base diode is the use of both junctions for one rectifying action which makes the collector-base diode unavailable for further use in the circuit (as desirable in a TTL or DTL gate).*

Resistors. Resistors are formed from bulk regions of semiconductor material accentuated in length-to-width ratio. Although the resistor is usually defined by the base diffusion, emitter or collector regions may also be used. Representative sheet resistivities (ρ_s) for the three regions are:

Emitter: 5 to 20 ohms/sq
Base: 200 to 400 ohms/sq
Collector: 50 to 200 ohms/sq

*These logic configurations are discussed in Chap. 8.

TABLE 5-1   Representative Transistor Parameters

Parameter	Symbol	Value
Common-emitter gain................	β or h_{FE}	30 to 100
Gain-bandwidth product............	f_t or f_α	300 to 1,000 MHz
Collector resistance................	R_C	50 to 100 ohms
Emitter-base junction capacitance (at 1 volt).....................	C_{TE}	1 to 3 pf/mil² (usually 2 to 8 pf)
Collector-base junction capacitance (at 1 volt).....................	C_{TC}	0.1 to 0.2 pf/mil² (usually 1 to 4 pf)
Collector-substrate junction capacitance (at 1 volt).............	C_{TS}	Less than 0.1 pf/mil² (usually 2 to 5 pf)
Collector saturation voltage (at 10 ma).....................	$V_{CE(SAT)}$	0.4 volt
Collector-base breakdown voltage.....	BV_{CBO}	15 to 30 volts
Emitter-base breakdown voltage......	BV_{EBO}	4 to 8 volts
Collector-substrate breakdown voltage	BV_{CS}	Greater than 30 volts
Collector leakage current (at 25°C)....	I_{CBO}	Less than 25 na
Emitter leakage current (at 25°C).....	I_{EBO}	Less than 15 na

The resistance of bulk regions is given by

$$R = \rho_s \frac{l}{w} \tag{5-1}$$

where l = length of resistor

w = width of resistor

The upper limit on resistance depends upon the total surface area that one is willing to allot to the resistor, and the requirements on tolerance and temperature coefficient. The following considerations relate to the variables affecting resistance values:

1. For a given maximum surface area, reduction of the resistor width can provide significant increases in resistance value, but only at the expense of precision and yield. The maximum length that one may utilize for a given area is

$$l = \frac{A}{2w} \tag{5-2}$$

Thus $R = \rho_s A / 2w^2$ at the limit. Suppose that a 5 percent tolerance must be allowed for the diffusion process and that geometry control can be held to ± 0.1 mil (2.5 microns). Table 5-2 illustrates the tradeoffs involved in reduction of width to increase resistance. The following are assumed: $\rho_s = 400$ ohms/sq, $A = 50$ mils².

TABLE 5-2

w, mils	l, mils	R, ohms	Resistor tolerance, percent	Representative overall yield, percent
2.5	10	1,600	9	98
2.0	12.5	2,500	10	95
1.0	25	10,000	15	90
0.5	50	40,000	25	80
0.25	100	160,000	45	60

The preceding figures are hypothetical, but basically representative of present state-of-the-art capabilities.

2. For a given geometry, the resistance can be increased by increasing the sheet resistivity, but only at the expense of transistor characteristics and temperature coefficient. The sheet resistivity may be increased by lowering the concentration of impurities in the base region or by making the diffusion more shallow. The former will increase the base ohmic drop and the error in the location of the base-collector junction, and the latter will require a corresponding shallow emitter diffusion (and hence lower emitter efficiency). In either case, the transistor characteristics will suffer.

In addition to the above considerations, the higher-resistivity material will change value faster with temperature than the lower-resistivity material. This phenomenon results from the more critical dependence of resistivity on thermally generated carriers and on mobility variations for lightly doped material. Characteristic temperature coefficients of resistance vary from 200 to 1,000 ppm/°C (0.02 to 0.1 percent/°C).

Capacitors. Capacitors are formed by back-biased p-n junctions or parallel-plate silicon dioxide structures (the latter is illustrated in Fig. 5-1). The voltage and doping dependence of junction capacitors were described in Chap. 3. The highly doped emitter-base junction provides greatest capacitance per unit area (≈ 0.7 pf/mil^2 at 1-volt bias), but also the lowest breakdown voltage. Capacitance values for the collector base are somewhat lower (typically 0.2 pf/mil^2 at 1 volt) but with improved breakdown-voltage characteristics. In general, junction diode capacitors are of low quality (low Q) and are not suitable for resonant-circuit applications. Typically, useful values of capacitance are obtained by constructing large-area transistors (100 mil^2) and connecting the two junctions in parallel.

In epitaxial structures with buried n$^+$ layers, special capacitor structures can be formed with high capacitance per unit area. An oxide area above such a buried layer is opened prior to the p-type isolation diffusion. During this diffusion, the p-type diffusion front reaches the n$^+$ layer and is "stopped" by the high n concentration. It has the appearance of a deep base region with an especially high p-type surface concentration. The n^+ emitter diffusion then forms a junction with a high impurity concentration on both sides and a correspondingly high capacitance per unit area (typically 1 pf/mil² at 1 volt). The breakdown voltage is, however, quite low (≈ 3 volts).

The parallel-plate SiO₂ structure (MOS capacitor) is a high-Q component whose capacitance is independent of voltage. With dielectric thicknesses on the order of 500 to 1000 Å, capacitance values ranging from 0.2 to 0.4 pf/mil² are attainable, with breakdown voltages in excess of 30 volts. Limitations on the total area used result from the increasing probability of defects (pinholes in the oxide) that will lower the breakdown voltage or short the capacitor entirely.

Custom-design Considerations

The design of custom circuits in integrated-circuit form requires a new appreciation of the relative costs of the various electronic components. For custom-designed integrated circuits, as opposed to conventional-circuit design, inductors are nonexistent and a transistor is cheaper than a 5-kilohm resistor. The determining factors in cost are the surface area required and the dual factors of tolerance and performance as they affect the yield. Using the typical performance characteristics that have been outlined in this discussion, the following relative-cost comparison can be made (in arbitrary cost units):

Component	Cost
Transistor	3
Diode	2
4-kilohm resistor at ± 30%	3
4-kilohm resistor at ± 10%	12
50-pf capacitor at ± 30%	9

From this cost illustration, we might anticipate a preponderance of transistors and diodes in integrated-circuit designs. This design pattern has occurred to a great extent. Most digital-circuit designs use transistor-rich configurations that would be unthinkable in conventional-circuit design. The RTL gate, for example, uses a transistor for every input lead.

Parasitic Effects

A component, or a group of components, is isolated from the substrate, and hence from the remainder of the circuit, by means of a reverse-biased p-n junction. We have referred to this junction as the collector-substrate junction. The presence of the doped substrate and the collector-substrate junction makes possible several modes of secondary interaction between components ("parasitic effects"). Two important parasitic effects are:

1. Secondary transistor action
2. Capacitive intercomponent and intracomponent coupling

Secondary Transistor Action. An integrated-circuit transistor is constructed of four layers:

$$
\text{p-n-p transistor} \left\{ \begin{array}{l} \text{Emitter n}^+ \\ \text{Base p} \\ \text{Collector n} \\ \text{Substrate p} \end{array} \right\} \quad \text{n-p-n transistor}
$$

Although the substrate is incidental to the transistor, it appears possible for a p-n-p transistor to be formed by the usual base, collector, and substrate regions. In this secondary transistor, the normal base would act as the secondary emitter, the normal collector as the secondary base, and the substrate as the secondary collector. In many structures, such transistor action is indeed possible, although with considerably degraded characteristic parameters. One would anticipate a low gain for the secondary transistor because of the low impurity concentration in the base (secondary emitter), causing a low emitter efficiency, and the width of the collector (secondary base), causing a low base-transport factor.

Let us consider the transistor in Fig. 5-1 as an example. Assume that the diffusion length for electrons in the p-type base is $L_n = 10$ microns, and that for holes in the n collector it is $L_p = 5$ microns. The emitter efficiency is Eq. (3-24), amended for the specific case:

$$
\gamma \approx \frac{1}{1 + D_n N_D L_p / D_p N_A w} = \frac{1}{1 + p_{\text{base}} L_p / p_{\text{base}} w_{\text{base}}}
$$

$$
= \frac{1}{1 + (0.1)(5)/(0.5)(1)} = 0.5 \tag{5-3}
$$

The base-transport factor is

$$
\alpha' = \frac{1}{1 + w^2_{\text{collector}}/2L_p^2} = \frac{1}{1 + (7)^2/2(5)^2} \approx 0.5 \tag{5-4}
$$

Thus the total common-base gain is

$$
\alpha = \gamma \alpha' = 0.25 \tag{5-5}
$$

giving a common-emitter gain of

$$\beta = \frac{\alpha}{1 - \alpha} \approx 0.33 \qquad (5\text{-}6)$$

For the case assumed, therefore, parasitic currents of nonnegligible values can flow through the secondary transistor. Secondary transistor β's as high as 0.4 to 0.6 have been reported. In such cases, the circuit may not behave as intended.

Capacitive Coupling. The collector-substrate capacitance, although very low in value beneath the collector, increases along the sidewalls toward the surface. Near the surface, the concentration of carriers is high, as dictated by the diffusion equation. As a result, collectors of adjacent transistors may be capacitively coupled to one another through the substrate by capacitances on the order of 5 pf or so. At 20 MHz, the impedance offered by this capacity is around 1,000 ohms, and significant cross-coupling can result.

Another aspect that must be considered is the shunt capacitive coupling on large-area resistors. An equivalent resistor circuit that takes into account the capacitive coupling is shown in Fig. 5-3. A 5-kilohm resistor with an area of 25 mil^2 and an associated coupling capacitance of 0.2 pf/mil^2 will experience a serious impedance drop-off and transmission loss at frequencies above 5 MHz. On the other hand, the phase shift due to this

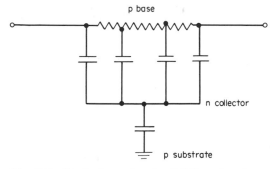

Fig. 5-3 Equivalent circuit of diffused resistor.

parasitic effect may often be used to advantage, e.g., in the construction of efficient phase-shift oscillators.

MOSFET Integrated Circuits

In MOSFET integrated circuits, the transistor often plays the role of all required circuit components. Because it has a particularly simple construction, the MOSFET integrated circuit allows functions of high circuit complexity.

The structure of an enhancement-mode MOSFET in an integrated circuit is illustrated in Fig. 5-4. Three devices are shown. Note that both the source and the drain are formed during the same p⁺ diffusion. No isolation diffusion between transistors is required since there is no preexisting surface channel associated with enhancement-mode devices. Depletion-mode devices do require, however, a "channel stopping" diffusion between devices for isolation.

The principal construction difficulties faced by early developers of practical devices were instabilities in the oxide characteristics and realization of required small-channel geometries. These problems have been largely overcome, and MOSFETs have been employed with dramatic effect in the integration of large blocks of circuitry.

We now consider prevalent characteristics of MOSFET components.

Transistors.   The interface between the silicon and silicon dioxide is extremely important in determining the threshold voltage of MOSFETs. In present-day devices, the surface states at the boundary encourage the accumulation of electrons.   Thus n-channel devices tend toward depletion-mode while p-channel devices are commonly enhancement-mode.   n-channel enhancement-mode and p-channel depletion-mode devices can be made by using a more heavily doped substrate, but only at the expense of the drain-to-substrate breakdown voltage.

The output voltage swing of enhancement-mode devices is dc-compatible with the input voltage requirements of similar devices. This characteristic is an important advantage in digital-circuit schemes, and enhancement-mode devices are used almost exclusively for such applications. On the other hand, the quiescently conducting characteristics of depletion-mode devices make them attractive for linear-circuit applications.

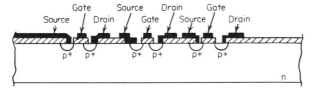

Fig. 5-4   MOSFET integrated-circuit structure.

Parameter values of MOSFETs are quite variable because of the high degree of correlation among parameters and the different proprietary advantages of the various manufacturers. A typical transistor might have the following characteristics:

TABLE 5-3

Parameter	Symbol	Value
Transconductance..................................	g_m	2,000 mhos
Threshold voltage (if enhancement-mode).............	V_{th}	3 volts
Pinch-off voltage (if depletion-mode)................	V_{po}	3 volts
Gate-to-source breakdown voltage...................	BV_{GS}	25 volts
Drain-to-source breakdown voltage..................	BV_{DS}	25 volts
Input capacitance..................................	C_g	5 pf
Output capacitance.................................	C_O	2 pf
Input resistance...................................	R_g	10^{15} ohms

Resistors. The resistance function in MOSFET integrated circuits is obtained by employing the channel conductance of a suitably biased transistor. A common configuration for enhancement-mode types is shown in Fig. 5-5. An n-channel device is assumed for the convenience of dealing with positive signs for the voltage quantities.

The gate and drain terminals are connected. Suppose V_{DS} starts at 0 volt and is increased in the positive direction. No current will flow until $V_{DS} = V_{th}$. As the voltage is increased beyond V_{th}, conduction characteristics of a

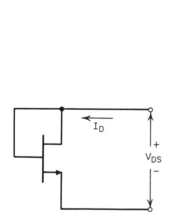

Fig. 5-5   Resistance configuration for enhancement-mode MOSFET.

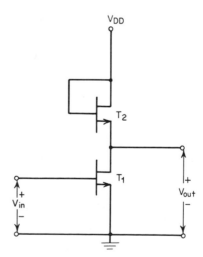

Fig. 5-6   MOSFET switch configuration.

saturated transistor will apply [Eq. (3-53)]. The transistor is in the saturated region of operation because $V_{DS} > V_{GS} - V_{th}$, since $V_{DS} = V_{GS}$. The current-voltage characteristics are therefore

$$I_D = 0 \qquad V_{DS} < V_{th}$$

$$I_D = \frac{\mu_n C_g}{2l^2} (V_{DS} - V_{th})^2 \qquad V_{DS} \geq V_{th} \tag{5-7}$$

The resistance is clearly nonlinear, but is useful for many applications. An example is the switch configuration of Fig. 5-6. Transistor T_2 is used as a load resistor. The channel of T_1 is made considerably wider than that of T_2, but the other parameters (including the channel length) are unchanged. Under these conditions, the output voltage will decrease linearly with increasing input voltage from a value near V_{DD} to a value near ground. The voltage gain in the transition region is equal to the square root of the width ratio, and is independent of the absolute transconductance values (see Prob. 5-1).

Capacitors. Capacitance requirements are normally met by utilizing the gate-to-source or other terminal capacitance. In applications where these are not sufficiently large, special MOS or junction capacitors can be constructed by methods already discussed. Since MOSFET circuitry is high-impedance, smaller capacitor values are required than in bipolar transistor circuitry.

PROBLEMS

5-1. Assume that transistors T_1 and T_2 of Fig. 5-6 are identical except for the respective values of gate capacitance, C_1 and C_2.

    a. Assume that a voltmeter (with a high but finite input resistance) is placed between the output terminal and ground. Show that $V_{out} = V_{DD} - V_{th}$ for $0 \leq V_{in} < V_{th}$.

    b. Discuss the operating mode of transistor T_1 for $V_{in} > V_{th}$. In particular, show that T_1 is saturated for

$$V_{th} \leq V_{in} < \frac{(C_2/C_1)^{1/2} V_{DD} + V_{th}}{1 + (C_2/C_1)^{1/2}}$$

    and unsaturated for greater values of V_{in}.

    c. Show that

$$V_{out} = V_{DD} - V_{th} - \left(\frac{C_1}{C_2}\right)^{1/2} (V_{in} - V_{th})$$

    for T_1 in the saturated region of operation. Note that the voltage gain dV_{out}/dV_{in} is proportional to the square root of the ratio of the channel width of T_1 to that of T_2.

d. Assume $C_1 = 9\ C_2$ and $V_{th} = 0.2\ V_{DD}$. Sketch V_{out} versus V_{in} for

$$0 \le V_{\text{in}} < \frac{(C_2/C_1)^{1/2}\ V_{DD} + V_{th}}{1 + (C_2/C_1)^{1/2}}$$

SIX

Digital-design Concepts...
Combinatorial-design Principles

Digital systems may be characterized as processors of discrete information, usually in the form of bivalued voltage levels. Digital circuitry is employed not only in computers but also in a wealth of processing, control, and communication equipment. When dealing with variables that can assume only two distinct values, the information content associated with any given processing stage is necessarily small. Thus, a large number of repetitive circuits are required for even modest functions. The circuit design can, however, be simpler and less critical than that for analog circuits, which deal with variables assuming a continuum of values. These two factors — large numbers of repetitive circuits and simplicity of design accentuate the attractiveness of integrated circuits for digital applications. It is not surprising that the early integrated circuits were digital designs, and that the majority of integrated circuits today are used in digital applications.

The study of the digital applications of integrated circuits will begin with a review of logical-design concepts common to all digital-design activities. Here we will discuss Boolean algebra, the formation of logical statements, the manipulation and minimization of design expressions, and the basic synthesis of combinatorial circuitry. The next chapter will consider the synthesis of sequential circuitry.

Boolean Algebra

Boolean algebra is a self-consistent mathematical development that is particularly applicable to logical rules of thought. It is based on a primary set of definitions and a number of postulates or axioms. The variables in the algebra can assume only one of two values, called "truth values." Thus a proposition is either true (truth value of 1) or not true (truth value of 0). For example, a typical proposition might take the following form:

Let R represent the proposition, "It is raining." Then "not R" or $\overline{R}$ represents the proposition, "It is not raining." The variable R assumes a value corresponding to the truth value of the statement, illustrated as follows:

State of weather	R	$\overline{R}$
Raining...............	1	0
Not raining............	0	1

Since Boolean algebra deals with variables that are limited to two discrete values, it is useful to introduce the binary number system and arithmetic operations in this system. Note carefully, however, that binary arithmetic and Boolean algebra are two different mathematical systems, and that operations within this system are defined differently.

Binary Number System. The binary number system is distinguished by the use of the base (or radix) 2 in representing numbers. Common decimal numbers use the base 10. For example, in expressing the decimal number 149, it is understood that we mean

$$149 = 1 \times 10^2 + 4 \times 10^1 + 9 \times 10^0$$

The allowed number symbols in any number system run from zero up to the number preceding the radix. In binary notation, the above number is

$$(149)_{10} = (10010101)_2 = 1 \times 2^7 + 0 \times 2^6 + 0 \times 2^5 \\ + 1 \times 2^4 + 0 \times 2^3 + 1 \times 2^2 + 0 \times 2^1 + 1 \times 2^0$$

The conversion of a binary number to a decimal number may be performed simply by rewriting each term in decimal form and summing. Thus, to convert $(10010101)_2$ to decimal form,

$$1 \times 2^7 = 128$$
$$0 \times 2^6 = 0$$
$$0 \times 2^5 = 0$$
$$1 \times 2^4 = 16$$
$$0 \times 2^3 = 0$$
$$1 \times 2^2 = 4$$
$$0 \times 2^1 = 0$$
$$1 \times 2^0 = \underline{1}$$
$$149$$

The conversion of a decimal number to a binary number may be performed by successive divisions by 2. The remainder terms from each division are the terms of the binary expression, starting with the least significant term from the first division. Thus, to convert $(149)_{10}$ to binary form,

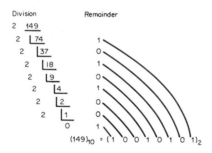

Division   Remainder

2 | 149
2 | 74   1
2 | 37   0
2 | 18   1
2 | 9   0
2 | 4   1
2 | 2   0
2 | 1   0
0   1

$$(149)_{10} = (1\ 0\ 0\ 1\ 0\ 1\ 0\ 1)_2$$

Arithmetic operations with binary numbers follow ordinary rules, taking into account the radix difference. Thus, when adding, a carry is generated when the number equals or exceeds the radix. To add $(10010101)_2$ and $(11011)_2 = (27)_{10}$, we proceed as follows:

$$1\ 0\ 0\ 1\ 0\ 1\ 0\ 1$$
$$1\ 1\ 0\ 1\ 1$$

① 0	Partial sum = 10, carry 1
① 0	Partial sum = 10, carry 1
① 0	Partial sum = 10, carry 1
① 0	Partial sum = 10, carry 1
① 1	Partial sum = 11, carry 1
1	Partial sum = 01, no carry
0	Partial sum = 00, no carry
1	Partial sum = 01, no carry

$$(1\ 0\ 1\ 1\ 0\ 0\ 0\ 0)_2 = (176)_{10}$$

Similar rules apply for other operations.*

*The interested reader is referred to any basic text on logical design, for example, Ref. 3, pp. 42–65.

Huntington's Postulates. In establishing the rules of Boolean algebra, we first assume the acceptance of common definitions, i.e., the meaning of equivalence, bracketed expressions, etc. The basis of Boolean algebra then follows from Huntington's postulates:

> Given the variables $A, B, C, \ldots$ which can take on the values 0 or 1 only, the following is postulated [NOTE: These postulates define the operations "+" (addition or "OR" function) and "·" (product or "AND" function)]:

Ia	$A + B = 0$ or 1	(closure under addition)
b	$A \cdot B = 0$ or 1	(closure under multiplication)
IIa	$A + 0 = A$	(existence of null element)
b	$A \cdot 1 = A$	(existence of identity element)
IIIa	$A + B = B + A$	(commutivity of addition)
b	$A \cdot B = B \cdot A$	(commutivity of multiplication)
IVa	$A + (B \cdot C) = (A + B) \cdot (A + C)$	
	(distributivity of addition)	
b	$A \cdot (B + C) = A \cdot B + A \cdot C$	
	(distributivity of multiplication)	
Va	$A + \bar{A} = 1$	
b	$A \cdot \bar{A} = 0$	

Postulates I, IVa, Va, and Vb do not follow regular arithmetic rules, and thus give Boolean algebra its special character. Note also that each postulate involving addition is closely related to another involving multiplication. These relations are called "duals." It is not surprising that the theorems developed from these postulates (see below) retain this dual nature. The expression $\bar{A}$ may be called "not A," "A-complement," or "A-bar."

Important Theorems. Several theorems of Boolean algebra that are of special importance are listed here:

1a	$A \cdot A = A$	
b	$A + A = A$	
2	$\bar{\bar{A}} = A$	
3a	$A + 1 = 1$	
b	$A \cdot 0 = 0$	
4a	$0 + 0 = 0$	
b	$1 \cdot 1 = 1$	
5a	$1 + 0 = 1$	
b	$1 \cdot 0 = 0$	
6a	$\bar{0} = 1$	
b	$\bar{1} = 0$	
7a	$A(A + B) = A$	
b	$A + AB = A$	

$$8a \quad A(\bar{A} + B) = AB$$
$$b \quad A + \bar{A}B = A + B$$
$$9a \quad \bar{A}(A + B) = \bar{A}B$$
$$b \quad \bar{A} + AB = \bar{A} + B$$

Note that by using Ths. 3a to 5b the following addition and multiplication tables may be constructed:

Addition	Multiplication
$0 + 0 = 0$	$0 \cdot 0 = 0$
$0 + 1 = 1$	$0 \cdot 1 = 0$
$1 + 0 = 1$	$1 \cdot 0 = 0$
$1 + 1 = 1$	$1 \cdot 1 = 1$

Combinatorial-design Principles

The principles of Boolean algebra just discussed find application in the formulation of design equations that bear a one-to-one correspondence with physical variables or events. Such formulation may be undertaken at an abstract level without regard to physical implementation by electronic circuitry. When we come to physical implementation, we must relate the variables or events to measurable quantities, such as voltage levels, and make the necessary decisions by means of switching circuits. Decisions that are based on the logical states or truth values of variables at any given time and are independent of the previous history of these variables are implemented by combinatorial circuitry.

Logical Statements. A logical statement indicates an outcome according to the states of independent variables or deciding factors. The following are logical statements with OR connectives:

"I will buy the suit if it is inexpensive OR if it is conservative."
"The train will stop if it needs fuel OR if a passenger wants to get off."

In these sentences, a given outcome (buy suit, train stop) is predicated upon one or another event occurring. An assignment of variables can be made to express the sentences in algebraic form. For example, in the second sentence, let

T = train stops   ($T = 1$ if train stops, $T = 0$ if not)
F = needs fuel   ($F = 1$ if train needs fuel, $F = 0$ if not)
P = passenger wants to get off   ($P = 1$ if passenger wants to get off, $P = 0$ if not)

Then

$$T = F + P \tag{6-1}$$

is an algebraic equation representing the logical statement. The symbol "+" replaces the connective "OR." If we wish to verify that the rules of addition just discussed provide the outcome required by the statement, we may do so by considering every possible combination of the variables F and P, and the corresponding outcome T for each case, in a table (truth table):

F	P	T
0	0	0
0	1	1
1	0	1
1	1	1

Next, consider the following logical statement with an AND connective:

"I will buy the suit if it is inexpensive AND conservative."

Here, a given outcome depends upon two events occurring simultaneously. Letting

B = buy suit
I = inexpensive
C = conservative

we have

$$B = I \cdot C \text{ (or just } IC) \tag{6-2}$$

The symbol "·" replaces the connective "AND." Again we may verify that the rules of multiplication coincide with the required statement by developing a truth table:

I	C	B
0	0	0
0	1	0
1	0	0
1	1	1

Alternatively, if we are presented the truth table as initial information, we may derive the appropriate equation for B by considering every combination of I and C that makes B true. This occurs in the preceding table when $I = 1$ and $C = 1$. We write

B is true if I is true AND C is true

or

$B = IC$

Let us try this procedure with the previous table for the OR connective. There are three combinations that make $T = 1$.

> T is true if F is false AND P is true
> OR F is true AND P is false
> OR F is true AND P is true

or

$$T = \overline{F}P + F\overline{P} + FP \tag{6-3}$$

Equation (6-3) may be simplified by use of the axioms and theorems listed in the previous section. We have

$$
\begin{aligned}
T &= \overline{F}P + F\overline{P} + FP + FP & \text{(Th. 1}b\text{)} \\
&= (\overline{F}P + FP) + (F\overline{P} + FP) & \text{(Ax. III}a\text{)} \\
&= (P\overline{F} + PF) + (F\overline{P} + FP) & \text{(Ax. III}b\text{)} \\
&= P(\overline{F} + F) + F(\overline{P} + P) & \text{(Ax. IV}b\text{)} \\
&= P \cdot 1 + F \cdot 1 & \text{(Ax. V}a\text{)} \\
&= P + F & \text{(Ax. II}a\text{)}
\end{aligned}
$$

which is our desired OR expression.

It is always desirable to simplify a Boolean expression to the greatest extent possible. This makes the relationship simpler and clearer and (in terms of electronic hardware) saves money. The algebraic approach, however, is often too tedious and uncertain to be economical as a design procedure. We shall therefore substitute for this approach a graphic technique that is easier. We must first discuss logical elements.

Logical Elements. The logical elements are diagrammatic symbols that represent various Boolean operations, such as the connectives. They may be considered as processors that accept one or more inputs and deliver the appropriate output. Later we will discuss circuits that perform the desired operations.

The first logical element is the inverter, which performs the NOT function. The symbol and function are indicated in Fig. 6-1. The inverter

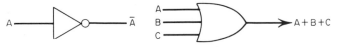

Fig. 6-1 Inverter.   **Fig. 6-2   OR gate.**

accepts a variable, let us say A, and provides "not A" ($\overline{A}$) at the output. Thus, if A is 1, the output will be 0, and vice-versa.

The symbol and function for the OR gate are shown in Fig. 6-2. The OR gate accepts two or more inputs (three are shown, A, B, and C) and provides the sum of the inputs at the output ($A + B + C$).

Fig. 6-3   AND gate.

The symbol and function for the AND gate are shown in Fig. 6-3. The AND gate accepts two or more inputs (three are shown, $A, B,$ and C) and provides the product of the inputs at the output (ABC).

As an example of the use of the logical symbols, consider the previous OR problem involving a train stopping, Eq. (6-1). The solution $T = F + P$ is drawn in Fig. 6-4a. The equivalent but unsimplified expression (6-3) is drawn in Fig. 6-4b. Both circuits perform exactly the same function, but Fig. 6-4b requires considerably more circuitry than Fig. 6-4a. The economic advantages of simplifying Boolean expressions are thus made obvious. The graphic technique for simplification that we will rely on, the Karnaugh map, is next discussed.

Karnaugh Map.   Consider a problem involving two variables, A and B. There are four and only four possible combinations of these variables: $AB, \bar{A}B, \bar{A}\bar{B},$ and $A\bar{B}$. Since A and B must each be in some given state, the sum of all possible combinations must equal 1:

$$AB + \bar{A}B + \bar{A}\bar{B} + A\bar{B} = 1 \tag{6-4}$$

These combinations as a group define a "universe" with four regions.

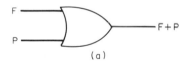

(a)

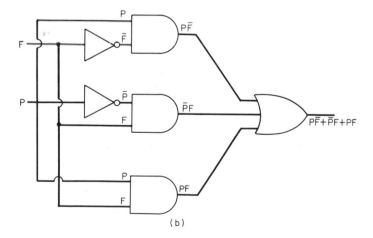

(b)

Fig. 6-4   Logical-circuit diagrams. (a) Simplified solution.
(b) Unsimplified solution for train problem.

We may draw this universe as a square with four boxes in it, as shown in Fig. 6-5. The two vertical columns represent the regions where A is not true and true, respectively. The two horizontal rows represent the regions where B is not true and true, respectively. The intersections of the columns and rows define the four states of $\overline{A}\overline{B}$, $A\overline{B}$, $\overline{A}B$, and AB.

An essential fact is that any two adjacent states in the figure may be combined with a net simplification. For example, suppose that we have the equation

$$T = \overline{A}\overline{B} + \overline{A}B \tag{6-5}$$

These are two adjacent states in the map. Note that they exactly coincide with the entire region where A is not true. Also, in going from one state to another, B changes value. The sum of the two states therefore does not depend on B; the important information is that A is not true. Thus we write

$$T = \overline{A} \tag{6-6}$$

This graphical simplification agrees with the algebraic method, where we would simplify Eq. (6-5) by the following steps:

$$\begin{aligned}
T &= \overline{A}\overline{B} + \overline{A}B \\
&= \overline{A}\ (\overline{B} + B) \\
&= \overline{A} \cdot 1 \\
&= \overline{A}
\end{aligned} \tag{6-7}$$

We may obtain a precise view of what is happening in the graphical presentation. We are combining two states that are identical except for one variable that changes states. We are thus factoring out $\overline{B} + B$.

Figure 6-5 is a two-variable Karnaugh map. We will draw the map in the manner shown in Fig. 6-6. The 0 and 1 represent the "not true" and "true" regions, respectively. For example, the state $\overline{A}\overline{B}$ is mapped into the upper left-hand box in Fig. 6-6.

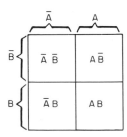

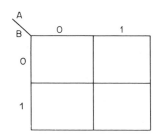

Fig. 6-5 "Universe" for two variables.

Fig. 6-6 Two-variable Karnaugh map.

Let us now use a Karnaugh map to simplify Eq. (6-3). A 1 is mapped into each of the states $\bar{F}P, F\bar{P},$ and FP, as shown in Fig. 6-7a. A 0 is entered in the remaining state. We then circle adjacent states in the map as shown in Fig. 6-7b. Note that the state FP is circled twice. This is allowed, and essentially it means that we are adding an extra FP term [which is what we did in the algebraic approach following Eq. (6-3)]. The two circled regions then represent P and F, and give $T = P + F$.

A Karnaugh map may be drawn for any number of variables. A three-variable map is illustrated in Fig. 6-8. We map combinations of BC in the four vertical columns and combinations of A in the two vertical rows. The sequence of states BC is carefully chosen so that any two adjacent boxes differ in only one variable. This is the single most important characteristic of all Karnaugh maps.

In the three-variable Karnaugh map, we look for adjacencies involving two or four boxes. One variable can be eliminated from an adjacency of two; two variables can be eliminated from an adjacency of four. For example, consider the expression

$$F = \bar{A}\bar{B}\bar{C} + \bar{A}\bar{B}C + \bar{A}BC + \bar{A}B\bar{C} + ABC + AB\bar{C} \qquad (6-8)$$

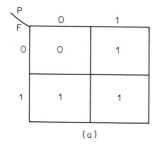

(a)

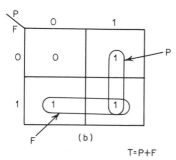

(b)

$T = P + F$

Fig. 6-7 *Two-variable Karnaugh map simplification. (a) Mapping states. (b) Circling adjacent states.*

BC A	00	01	11	10
0	$\bar{A}\bar{B}\bar{C}$	$\bar{A}\bar{B}C$	$\bar{A}BC$	$\bar{A}B\bar{C}$
1	$A\bar{B}\bar{C}$	$A\bar{B}C$	ABC	$AB\bar{C}$

Fig. 6-8 *Three-variable Karnaugh map.*

which is mapped in Fig. 6-9. Two circles of four can be drawn. Each circle has only one variable that remains the same throughout. We write

$$F = \bar{A} + B \tag{6-9}$$

As a final example of simplification using a three-variable Karnaugh map, consider the expression

$$T = \bar{A}\bar{B}\bar{C} + A\bar{B}\bar{C} + \bar{A}B\bar{C} + ABC \tag{6-10}$$

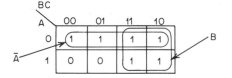

Fig. 6-9 Sample simplification of three-vari-
able map.

which is mapped in Fig. 6-10. There are two circles of two, and one which encompasses a term with no adjacencies. Note well that the boxes on either end of the map are adjacent. The term $\bar{A}\bar{C}$ is a "sneak adjacency"; watch out for those! The simplified function is

$$T = \bar{A}\bar{C} + \bar{B}\bar{C} + ABC \tag{6-11}$$

A four-variable Karnaugh map is shown in Fig. 6-11. The function

$$F = \bar{A}\bar{B}CD + \bar{A}\bar{B}C\bar{D} + \bar{A}B\bar{C}\bar{D} + \bar{A}B\bar{C}D + \bar{A}BCD$$
$$+ \bar{A}BC\bar{D} + AB\bar{C}D \tag{6-12}$$

is plotted in the map. Two circles of four and one circle of two can be drawn. The simplified function is

$$F = \bar{A}B + \bar{A}C + B\bar{C}D \tag{6-13}$$

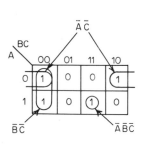

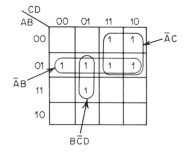

Fig. 6-10 Sample of
three-variable
map.

Fig. 6-11 Four-variable sample simplification.

In the four-variable map, we look for adjacencies of two, four, or eight. With an adjacency of eight, three variables can be eliminated. It should be apparent that the larger we can draw circles, the simpler the result — even if this requires redundant circling of terms. Two examples of sneak adjacencies are shown in Fig. 6-12a and b.

Combinatorial-design Examples

While the concepts that we have discussed thus far are simple, they have great usefulness and flexibility in design. Ease in formulating logical statements into design equations requires practice and some familiarity with "tricks of the trade." In this and the next sections, we will use combinatorial design to expand our working abilities.

   Two-point Lamp Control.   Let us first consider the problem of the control of a lamp L from either of two switches, A or B. This problem is analogous to the switching arrangement commonly found in households. The following assignment of variables is made:

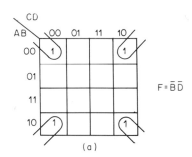

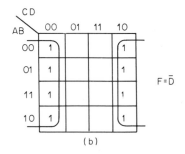

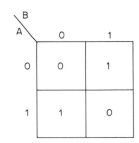

Fig. 6-12   Examples of sneak adjacencies in four-variable map.

Fig. 6-13   Karnaugh map for lamp problem.

$L = 1$ if lamp lit
$A = 1$ if switch A in "up" position
$B = 1$ if switch B in "up" position

The desired operation of the switching arrangement is indicated by the following truth table:

A	B	L
0	0	0
0	1	1
1	0	1
1	1	0

The function L is plotted in the Karnaugh map of Fig. 6-13. There are no adjacencies, and hence no simplification is possible. The resultant function is

$$L = \bar{A}B + A\bar{B} \tag{6-14}$$

Equation (6-14) is an important and commonly occurring relation which states that L will be true if either A or B is true, but not both. For this reason, Eq. (6-14) is called the "exclusive OR" of A and B, and is often abbreviated as follows:

$$\bar{A}B + A\bar{B} \equiv A \oplus B \tag{6-15}$$

It may be shown (Prob. 6-3) that

$$(A \oplus B) \oplus C = A \oplus (B \oplus C)$$
$$= \bar{A}B\bar{C} + A\bar{B}\bar{C} + \bar{A}\bar{B}C + ABC \tag{6-16}$$

Arithmetic Addition of Two Binary Numbers. Consider the problems of the arithmetic (as opposed to the logical) addition of two binary numbers, which for the sake of simplicity we take as being two digits in length each. The numbers A_1A_0 and B_1B_0 have a maximum possible sum of six, which requires a three-digit representation, $F_2F_1F_0$.

The problem may be attacked by considering the entire sum at one time or the digit-by-digit sum, together with the necessary carries into other digit positions. We will try both procedures.

If we consider the entire sum at one time, the desired operation may be written as in the following truth table:

A_1	A_0	B_1	B_0	F_2	F_1	F_0
0	0	0	0	0	0	0
0	0	0	1	0	0	1
0	0	1	0	0	1	0
0	0	1	1	0	1	1
0	1	0	0	0	0	1
0	1	0	1	0	1	0
0	1	1	0	0	1	1
0	1	1	1	1	0	0
1	0	0	0	0	1	0
1	0	0	1	0	1	1
1	0	1	0	1	0	0
1	0	1	1	1	0	1
1	1	0	0	0	1	1
1	1	0	1	1	0	0
1	1	1	0	1	0	1
1	1	1	1	1	1	0

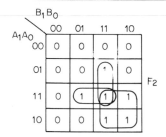

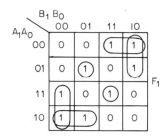

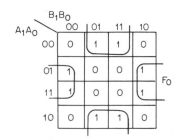

Fig. 6-14   Maps for simultaneous addition.

There are three outputs F_2, F_1, and F_0 which we seek to establish as functions of the four inputs A_1, A_0, B_1, and B_0. We will need a Karnaugh map for each output. The three maps are shown in Fig. 6-14.

We obtain from the maps the following equations for our adder:

$$F_2 = A_1B_1 + A_0B_1B_0 + A_1A_0B_0$$
$$F_1 = \bar{A}_1\bar{A}_0B_1 + \bar{A}_1B_1\bar{B}_0 + A_1\bar{B}_1\bar{B}_0 + A_1\bar{A}_0\bar{B}_1$$
$$+ \bar{A}_1A_0\bar{B}_1B_0 + A_1A_0B_1B_0 \quad (6\text{-}17)$$
$$F_0 = \bar{A}_0B_0 + A_0\bar{B}_0 = A_0 \oplus B_0$$

This procedure is straightforward but gets quite complicated as the number of digits to be added is increased. To obtain the design equations for the addition of two three-digit numbers would require a truth table with 64 entries; two four-digit numbers 256 entries; etc.

The second procedure for adding that we will discuss remains simple for an arbitrary-length adder and allows simple extension of the number of digits added without complete redesign. Here we consider just one stage of addition, the nth digits. The inputs to the stage are A_n, B_n, and C_{n-1}, the last term being the carry from the previous stage. The outputs are S_n (the nth digit sum) and C_n (the carry to the next stage). The operation of the stage is indicated in the truth table that follows:

A_n	B_n	C_{n-1}	S_n	C_n
0	0	0	0	0
0	0	1	1	0
0	1	0	1	0
0	1	1	0	1
1	0	0	1	0
1	0	1	0	1
1	1	0	0	1
1	1	1	1	1

We next map the functions S_n and C_n in the Karnaugh maps of Fig. 6-15. We obtain

$$S_n = \bar{A}_n\bar{B}_nC_{n-1} + \bar{A}_nB_n\bar{C}_{n-1} + A_nB_nC_{n-1} + A_n\bar{B}_n\bar{C}_{n-1} \quad (6\text{-}18)$$

$$C_n = A_nB_n + B_nC_{n-1} + A_nC_{n-1} \quad (6\text{-}19)$$

By comparison to Eq. (6-16), Eq. (6-18) may be rewritten

$$S_n = A_n \oplus B_n \oplus C_{n-1} \quad (6\text{-}20)$$

Thus the nth sum digit may be thought of as being formed by two partial sums, effected with "exclusive OR" operations. For this reason, the "exclusive OR" function is often called the half adder or the mod 2 adder. A stage-by-stage design, where one deals with a typical nth stage and employs the same design for every stage, is often called an iterative design.

Selection Network. As a final example, consider the problem of activating any one of several output lines by means of a binary-coded input. To be specific, suppose there are three input lines A_2, A_1, A_0 which can represent in binary code the numbers 0 through 7. We wish to activate any one of eight output lines F_0, F_1, . . . , F_7 such that the line activated will correspond to the number $A_2 A_1 A_0$. A diagram of the selection network is shown in Fig. 6-16.

The operation of the selection network is indicated in the truth table that follows:

A_2	A_1	A_0	F_7	F_6	F_5	F_4	F_3	F_2	F_1	F_0
0	0	0	0	0	0	0	0	0	0	1
0	0	1	0	0	0	0	0	0	1	0
0	1	0	0	0	0	0	0	1	0	0
0	1	1	0	0	0	0	1	0	0	0
1	0	0	0	0	0	1	0	0	0	0
1	0	1	0	0	1	0	0	0	0	0
1	1	0	0	1	0	0	0	0	0	0
1	1	1	1	0	0	0	0	0	0	0

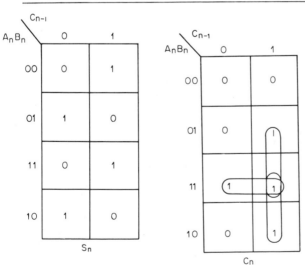

Fig. 6-15 *Maps for modular addition.*

The equations for the selection network are

$$F_0 = \bar{A}_2\bar{A}_1\bar{A}_0$$
$$F_1 = \bar{A}_2\bar{A}_1 A_0$$
$$F_2 = \bar{A}_2 A_1\bar{A}_0$$
$$F_3 = \bar{A}_2 A_1 A_0$$
$$F_4 = A_2\bar{A}_1\bar{A}_0$$
$$F_5 = A_2\bar{A}_1 A_0$$
$$F_6 = A_2 A_1\bar{A}_0$$
$$F_7 = A_2 A_1 A_0$$

(6-21)

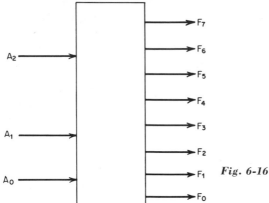

Fig. 6-16   *Selection network.*

Additional Design Tools

Excluded States and "Don't Care" Conditions.   Very often in practical design situations, one or more combinations of variables can never occur. These variable conditions, called "excluded states," may be used to simplify design equations.

As an example application of excluded states, suppose we know that in the previous selection network the states of $A_2 A_1 A_0$ will never occur as 000, 101, 110, and 111 (i.e., we will never need to activate lines F_0, F_5, F_6, or F_7). The abbreviated truth table follows:

A_2	A_1	A_0	F_4	F_3	F_2	F_1
0	0	1	0	0	0	1
0	1	0	0	0	1	0
0	1	1	0	1	0	0
1	0	0	1	0	0	0

Let us employ Karnaugh maps for the functions F_4, F_3, F_2, and F_1, placing a $\times$ in those $A_2 A_1 A_0$ states that are excluded. These are shown in Fig. 6-17.

Since the excluded states will never occur, we may freely treat them as 1s or 0s in the maps, according to their usefulness in permitting simplification of the expressions. Thus, in Karnaugh maps, we may circle those $\times$'s that result in adjacencies to the 1s. The resulting simplified equations are

$$
\begin{aligned}
F_1 &= \bar{A}_2 \bar{A}_1 \\
F_2 &= A_1 \bar{A}_0 \\
F_3 &= A_1 A_0 \\
F_4 &= A_2
\end{aligned}
\tag{6-22}
$$

Another form of extraneous state is the "don't care" condition. Suppose that we definitely want F_0 to be activated when $A_2 A_1 A_0 = 000$, but "don't care" whether or not it is activated when $A_2 A_1 A_0 = 001, 010$, or 111. The Karnaugh map for F_0 may then be drawn as shown in Fig. 6-18, with the resulting simplified equation

$$
F_0 = \bar{A}_2 \bar{A}_1 \qquad (\text{or } \bar{A}_2 \bar{A}_0)
\tag{6-23}
$$

Complementing Equations. We have defined the NOT function (or complement) of a variable, and have seen the need for complemented variables in many design equations.

Suppose, however, that the variable to be complemented is a function of

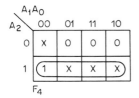

F_4

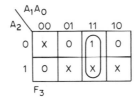

F_3

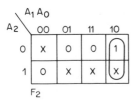

F_2

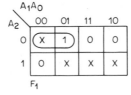

F_1

Fig. 6-17 *Sample simplification using excluded states.*

other variables, and we wish to express the complemented variable in terms of the other variables. As an example, suppose

$$F_2 = \bar{A}\bar{B}\bar{F}_0$$
$$F_0 = AC + \bar{B}\bar{C} + \bar{A}\bar{C}$$

(6-24)

and we wish to find F_2 as a function of $A, B,$ and C. We have

$$F_2 = \bar{A}\bar{B}\overline{(AC + \bar{B}\bar{C} + \bar{A}\bar{C})}$$

(6-25)

We have not, to this point, considered how to find the complement of an expression such as the one in parentheses in Eq. (6-25). Such expressions may be complemented by means of De Morgan's theorem, which states

$$\overline{X + Y} = \bar{X}\bar{Y}$$

(6-26)

$$\overline{XY} = \bar{X} + \bar{Y}$$

(6-27)

In essence, then, in order to complement an expression, we must change addition operations to product operations, product operations to addition operations, and variables to their complements.

Returning to Eq. (6-24), we have

$$\bar{F}_0 = \overline{(AC + \bar{B}\bar{C} + \bar{A}\bar{C})}$$
$$= (\overline{AC})(\overline{\bar{B}\bar{C}})(\overline{\bar{A}\bar{C}})$$
$$= (\bar{A} + \bar{C})(B + C)(A + C)$$
$$= (\bar{A}B + \bar{A}C + B\bar{C} + C\bar{C})(A + C)$$
$$= A\bar{A}B + A\bar{A}C + AB\bar{C} + AC\bar{C} + \bar{A}BC + \bar{A}CC$$
$$\quad + B\bar{C}C + CC\bar{C}$$
$$\bar{F}_0 = AB\bar{C} + \bar{A}BC + \bar{A}C$$

(6-28)

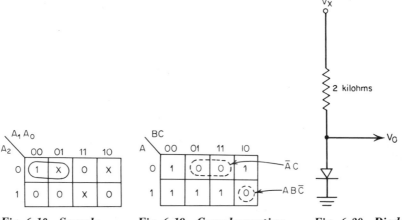

Fig. 6-18   Sample simplification using "don't care" conditions.

Fig. 6-19   Complementing by Karnaugh map.

Fig. 6-20   Diode circuit.

Equation (6-28) can be further simplified (by Karnaugh map) to

$$\overline{F}_0 = AB\overline{C} + \overline{A}C \tag{6-29}$$

Let us plot F_0 (not complemented) in the Karnaugh map of Fig. 6-19.

Note that if we circle the 0s instead of the 1s, we obtain $\overline{A}C + AB\overline{C}$ which is exactly $\overline{F}_0$. This result must be true since $F_0 + \overline{F}_0 = 1$. Thus, a Karnaugh map may be used to determine the complement of an expression by circling the 0s in the map.

The final result of our original problem, Eq. (6-25), is

$$
\begin{aligned}
F_2 &= \overline{A}\,\overline{B}(\overline{A}C + AB\overline{C}) \\
&= \overline{A}\,\overline{B}C + A\overline{A}B\overline{B}\overline{C} \\
&= \overline{A}\,\overline{B}C
\end{aligned} \tag{6-30}
$$

Physical Implementation

In this section, we discuss switching circuits that perform the logical functions with which we have gained some familiarity — AND, OR, and inverter. We will discover that judicious factoring of logical equations, beyond those steps assisted by the Karnaugh map, can save hardware at the expense of electrical performance.

For a number of reasons, it has been found to be advantageous to use functions other than the AND, OR, and inverter as primary building blocks in design. These functions, the NAND and NOR, are discussed later in this chapter.

Switching Circuitry. To appreciate the operation of switching circuits, we have to briefly review the electrical characteristics of their principal components, the diode and the transistor. Consider the circuit of Fig. 6-20, composed of a battery, a resistor, and a diode. We are interested in the output voltage V_O for positive and negative values of V_X. Suppose $V_X = +5$ volts. The diode is biased in the forward direction and is conducting. Ideally, there will be no voltage drop across the diode; in practice we know that there will be a small voltage drop, let us say 0.5 volt. In this case, then, $V_O = +0.5$ volt. The current through the resistor is

$$I = \frac{V_X - V_O}{R} = \frac{4.5 \text{ volts}}{2 \times 10^3 \text{ ohms}} = 2.25 \times 10^{-3} \text{ amp} \tag{6-31}$$

Now suppose that $V_X = -5$ volts. The diode is back-biased and does not conduct. In the absence of another circuit path to ground, no current flows through the resistor, and the voltage drop across the resistor is zero. As a consequence, $V_O = V_X = -5$ volts.

Next, consider the transistor circuit of Fig. 6-21. To make the transistor conduct, a small base voltage V_B is required. We assume $V_B = +0.5$ volt before conduction takes place. Suppose $V_X = 0$ volt. There will be no base current, and hence no collector current. Therefore $V_C = +5$ volts. Now suppose $V_X = +5$ volts, and that the β of the transistor is 20. The base current is

$$I_B = \frac{V_X - V_B}{R_1} = \frac{4.5 \text{ volts}}{10^4 \text{ ohms}} = 4.5 \times 10^{-4} \text{ amp} \tag{6-32}$$

In an ideal transistor, the collector current would be

$$I_C = I_B = 9 \times 10^{-3} \text{ amp} \tag{6-33}$$

and the voltage drop across R_2 would be

$$R_2 I_C = 18 \text{ volts} \tag{6-34}$$

giving

$$V_C = 5 - 18 = -13 \text{ volts} \tag{6-35}$$

We know, however, that the transistor saturates before the collector potential can be reversed. The collector saturation voltage is approximately

$$V_{C\text{SAT}} = +0.5 \text{ volt} \tag{6-36}$$

The actual collector current is therefore

$$I_C = \frac{4.5 \text{ volts}}{2 \text{ K}} = 2.25 \times 10^{-3} \text{ amp} \tag{6-37}$$

The base current is in excess of that required to achieve this collector current (four times as much). Note that a high value of V_X results in a low value of V_C and vice versa.

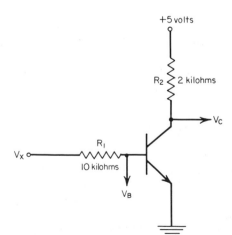

Fig. 6-21   Transistor circuit.

The AND Gate. A circuit that performs the AND function is shown in Fig. 6-22. Two inputs A and B are shown — others may be added in a similar manner. Let us make the following assignment of variables for A, B, and F:

Voltage range	Logical value
0–1.5 volts	0
1.5–3.0 volts	?
3.0–5.0 volts	1

From the previous discussion, if either A or B are at 0 volt, the output will be low in value, let us say 0.5 volt. If both A and B are at +5 volts, the output will be high (+5 volts). Two truth tables are drawn to describe the operation, one in terms of physical quantities, the other in corresponding logical values, as follows:

Voltage, volts			Logical value		
A	B	F	A	B	F
0	0	0.5	0	0	0
0	5	0.5	0	1	0
5	0	0.5	1	0	0
5	5	5	1	1	1

From the second of the two tables, we obtain

$$F = AB \tag{6-38}$$

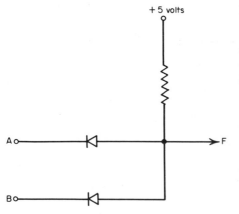

Fig. 6-22   AND gate.

which is the desired AND function. For more than two inputs, one obtains

$$F = ABC \cdots \tag{6-39}$$

The OR Gate. A circuit that performs the OR function is shown in Fig. 6-23. We define a correspondence between physical and logical variables as in the previous case. The physical and logical truth tables for the circuit follow:

Voltage, volts			Logical value		
A	B	F	A	B	F
0	0	0	0	0	0
0	5	4.5	0	1	1
5	0	4.5	1	0	1
5	5	4.5	1	1	1

The function is therefore

$$F = \bar{A}B + A\bar{B} + AB = A + B \tag{6-40}$$

The Inverter. An inverter circuit is basically that shown in Fig. 6-22. We have seen that a high voltage at the input (V_X) results in a low voltage at the output (V_C) and vice versa. If we associate the variables A with the input and F with the output, we have

$$F = \bar{A} \tag{6-41}$$

Factoring Equations. Associated with each output of an AND or OR gate is a diode. It is often desirable to factor out common terms from an

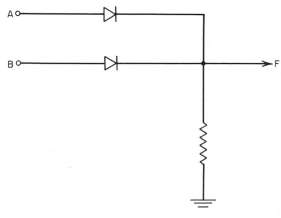

Fig. 6-23 OR gate.

expression in order to minimize the number of inputs required for each gate. In conventional-circuit design, this may be done to minimize the number of diodes, and hence cost. In integrated-circuit design, it is often necessary because of limitations in the number of inputs available in a given pur-chased circuit package.

There are no hard-and-fast rules for simplification that can be applied to an arbitrary case. One must be prepared to use a little ingenuity and a few trials to obtain the best solution (or at least a satisfactory solution).

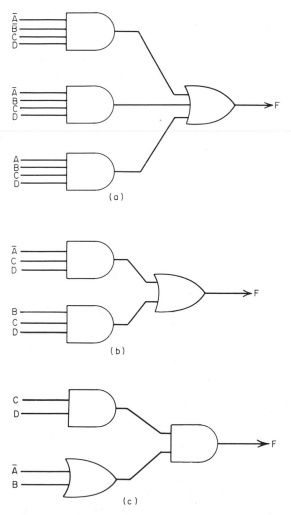

Fig. 6-24   Sample simplification. (a) Before simplification.
(b) Reduced two-level design. (c) Factored
design.

By way of example, suppose we wish to design circuitry which will achieve the following function:

$$F = \bar{A}\bar{B}CD + \bar{A}BCD + ABCD \tag{6-42}$$

Before Karnaugh-map simplification, we might be tempted to use the circuits of Fig. 6-24a. Exclusive of required inversion, Fig. 6-24a has the following part count:

    4 resistors
    15 diodes

The function of Eq. (6-43) may be plotted in a Karnaugh map and simplified to read

$$F = \bar{A}CD + BCD \tag{6-43}$$

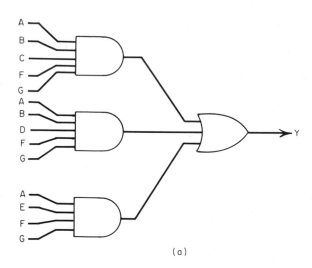

(a)

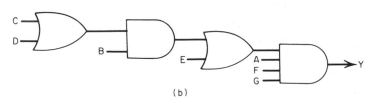

(b)

Fig. 6-25 *Implementation of equation* $Y = ABCFG + ABDFG$
$+ AEFG.$ *(a) Two level design. (b) Factored design.*

which corresponds to the circuit shown in Fig. 6-24b. The improved design has the following part count:

3 resistors
8 diodes

A further simplification may be achieved by factoring out the term CD from Eq. (6-44) to obtain

$$F = CD(\bar{A} + B) \tag{6-44}$$

A circuit corresponding to the factored expression is shown in Fig. 6-24c. The factored circuit has the following part count:

3 resistors
6 diodes

We thus see that judicious factoring of expressions can often simplify the circuitry required. Factoring often, however, increases the number of series stages through which a signal must propagate and hence may limit the speed of operation. For example, consider the following expression:

$$Y = ABCFG + ABDFG + AEFG \tag{6-45}$$

This function may be implemented by either of the circuits of Fig. 6-25.

The two-level (unfactored) design requires 17 diodes, but only two stages through which signals must propagate. The factored design requires only 10 diodes, but four stages through which the signals must propagate. The electrical specifications of the design will govern the extent of factoring which may be permitted.

Diode-gate Shortcomings. The use of circuitry that performs the same functions as the connectives in logical expressions is simple and straightforward. Nevertheless, the diode arrangements just discussed have certain disadvantages from the viewpoint of electronics that make them inconvenient to use in many applications. Two important drawbacks are:

1. Deleterious effects of loading
2. Loss of signal in transmission

Deleterious Effects of Loading. In many cases, the electrical design of diode gates will require specific information concerning the logical design. Thus uniform building blocks cannot be utilized for arbitrary designs.

Consider, for example, the connection shown in Fig. 6-26. An AND gate is shown feeding into one or more OR gates. Suppose we select $R_A = 2$ kilohms. What are the limitations on the value of R_O? Let us first take the case with one OR gate. We have previously agreed to interpret a voltage between 0 and 1.5 volts as a logical 0 and a voltage between 3.0 and 5.0 volts as a logical 1.

If the inputs A and B are both high, then the output F should be high. Considering the circuit of Fig. 6-26, an expression for the output voltage under these conditions is

$$V_F = \frac{5 \text{ volts} - 0.5 \text{ volt}}{R_A + R_O} R_O \geq 3.0 \text{ volts}$$

$$V_F = \frac{4.5}{2\text{K} + R_O} R_O \geq 3.0 \text{ volts}$$

(6-46)

Solving for R_O, we obtain

$$R_O \geq 4 \text{ kilohms}$$

(6-47)

We thus must limit the load on the AND gate in order to maintain the logical-1 output voltage. The situation deteriorates when the AND gate is

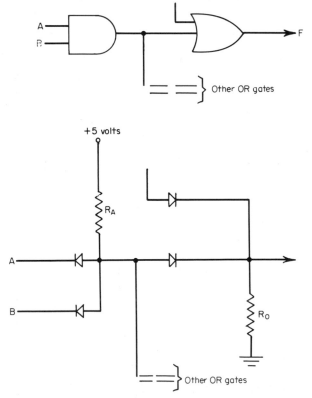

Fig. 6-26 *Connection illustrating loading problem.*

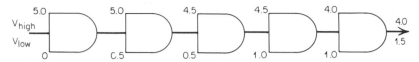

Fig. 6-27   Series array of gates.

connected to more than one OR gate. Consider the case of N OR gate loads. The expression for the output voltage with both A and B high is

$$V_F = \frac{4.5}{R_A + R_O/N} \frac{R_O}{N} \geq 3.0 \tag{6-48}$$

and

$$R_O \geq 2NR_A \tag{6-49}$$

For $N = 5$,

$$R_O \geq 10\,R_A = 20 \text{ kilohms} \tag{6-50}$$

If we were now to consider the opposite case, i.e., OR gate feeding into AND gates, we would find a similar requirement for R_A to be high in value with respect to R_O. Therefore, one gate-design type cannot satisfy arbitrary requirements for logical design.

Loss of Signal in Transmission.   If we have a number of diode gates in series, we would find that the logical-0 level rises from gate to gate due to the diode drops. The logical-1 level correspondingly falls. After the signal has propagated through a number of gates it loses its value. For example, consider the array of Fig. 6-27. We ignore the voltage-dividing effects of the resistors. The values of high- and low-level voltages that would appear at different stages are indicated in the figure. At the last stage we have reached the upper limit allowed on V_{low}.

A solution to both problems just discussed is to tie in an amplifier at the end of each gate. The amplifier acts to buffer the effects of loading and to reestablish the logic levels at the end of each logic stage. The simplest amplifier, however, is also an inverter. The combined OR gate-amplifier performs the function "not OR" or NOR, and the combined AND gate-amplifier the function "not AND" or NAND.

Inverting Gates.   NOR and NAND gates are the principal gating structures utilized in integrated-circuit design. Their extensive use stems from a number of advantages:

1. They are economical to fabricate as integrated circuits.

2. Standard gates may be used in arbitrary logical design.

3. Any logical equation may be implemented entirely with NAND gates or with NOR gates.

We now consider the logical properties of inverting gates and the special design procedures associated with these structures.

The NAND Gate. The symbol and function of the NAND gate are shown in Fig. 6-28. The single functions NOT, AND, and OR may be

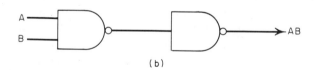

F = $\overline{ABCD}$ = $\overline{A}+\overline{B}+\overline{C}+\overline{D}$ *Fig. 6-28   NAND gate.*

implemented solely with NAND gates, as shown in Fig. 6-29a, b, and c. A direct replacement of the structures shown for diode gates, however, is not an economical design procedure. NAND gates may be used as primary building blocks in two-level designs by following the simple procedure outlined below:

1. Map function in Karnaugh map.

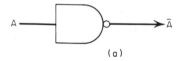

(a)

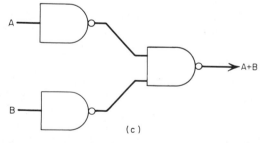

(b)

(c)

Fig. 6-29   Implementation of NOT, AND, and OR functions with NAND gates.

2. Minimize by circling 1s where permitted.

3. Draw block diagram in identical manner as in two-level AND/OR design, replacing both AND gates and OR gates with single NAND gates.

As an example of the above procedure, consider the expression

$$F = ABC + \bar{A}BC + \bar{A}\bar{B}C \tag{6-51}$$

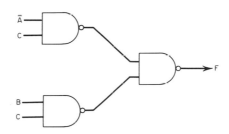

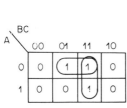

Fig. 6-30 *NAND gate design example.*   **Fig. 6-31** *Sample NAND design implementation.*

which is drawn in the Karnaugh map of Fig. 6-30. The simplified expression is

$$F = \bar{A}C + BC \tag{6-52}$$

which is implemented with the circuit in Fig. 6-31.

The NOR Gate.   The symbol and function of the NOR gate is shown in Fig. 6-32. A two-level design procedure for NOR gates is outlined here:

1. Map function in Karnaugh map.

2. Find minimized complement of desired expression by circling 0s.

3. Draw block diagram as in NAND design, but use as inputs the complements of terms in expression of step 2.

As an example of the above procedure, let us again consider the expression

$$F = ABC + \bar{A}BC + \bar{A}\bar{B}C \tag{6-51}$$

which is drawn in the Karnaugh map of Fig. 6-33.
We obtain

$$\bar{F} = \bar{C} + A\bar{B} \tag{6-53}$$

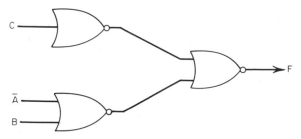

Fig. 6-32   NOR gate.   **Fig. 6-33   NOR gate design example.**

The NOR design for F is therefore the circuit of Fig. 6-34.

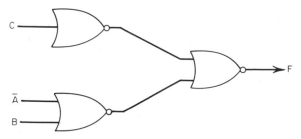

Fig. 6-34   Sample NOR design implementation.

REFERENCES

Digital-circuit Theory

1. Maley, G. A., and J. Earle: *The Logic Design of Transistor Digital Computers,* Prentice-Hall, 1963.
2. McCluskey, E. J.: *Introduction to the Theory of Switching Circuits,* McGraw-Hill, 1965.
3. Phister, Montgomery, Jr.: *Logical Design of Digital Computers,* John Wiley, 1958.

PROBLEMS

6-1. Steps in the proof of Th. 1a are outlined here. Justify the steps by reference to Huntington's postulates and prove the dual Th. 1b.

Theorem 1a $AA = A$

Proof:

$$A \cdot 1 = A$$
$$A(A + \bar{A}) = A$$
$$AA + A\bar{A} = A$$
$$AA + 0 = A$$
$$AA = A$$

6-2. Draw a block diagram of a circuit that performs the following functions:

$a.$ $F = A\bar{B} + \bar{A}B$

    b. $F = A + B + AC$

    c. $F = ABCD + \bar{A}B\bar{C}D$

Can any of the above expressions be simplified?

6-3. Verify Eq. (6-16).

6-4. Using a Karnaugh map, simplify the following expressions:

    a. $F = \bar{A}\bar{B}\bar{C} + \bar{A}B\bar{C} + ABC + A\bar{B}C$

    b. $F = \bar{A}B\bar{C} + AB\bar{C} + BC$

    c. $F = \bar{A}\bar{B}\bar{C} + \bar{A}\bar{B}D + \bar{A}\bar{B}C\bar{D} + \bar{A}B\bar{C}D + A\bar{B}C\bar{D} + AD$

6-5. A man is faced with the problem of crossing from the south to the north bank of a river with his dog, his goose, and a sack of corn. Unfortunately, his rowboat has room for only one passenger or piece of freight, so that the man must make several trips across the river. Normally his animals are well behaved and will wait unattended at the bank of the river if he so directs them. However animal instincts being what they are, in his absence and given the opportunity, the goose will eat the corn and the dog will eat the goose. Let the following variables be defined:

    $M = 1$ if man on north bank

    $G = 1$ if goose on north bank

    $D = 1$ if dog on north bank

    $C = 1$ if corn on north bank

Design a circuit that will indicate a tragic logistic situation.

6-6. Draw NAND and NOR designs for the following expressions:

    a. $L = M\bar{D}\bar{G} + M\bar{G}\bar{C} + \bar{M}DG + \bar{M}GC$

    b. $F = A\bar{B} + \bar{A}B$

    c. $F = A + B$

    d. $F = AB$

SEVEN

Sequential-design Principles

Combinatorial design derives logical functions that are based on real-time events. That is, a decision (the output) at any time t is based in a predictable manner on the inputs at that time t, minus perhaps the small increment of time required for the input signals to propagate through, and be processed by, the gate. Complete design flexibility requires some means for storing information at some time t so that it may affect a logical decision at a later time $t + \tau$. For example, suppose we wish to construct a device whose output will go to a logical 1 when all of its inputs are logical 1s and have been logical 1s for three previous timed increments. This design cannot be constructed solely with logical gating functions. What is needed is some information-storage function. This function is attained through use of the bistable or "flip-flop" circuit.

Bistable Circuits

Direct-coupled Set-reset Flip-flop. A simple bistable circuit may be constructed from two inverting gates of either the NAND or NOR variety. In Fig. 7-1, a direct-coupled set-reset flip-flop (DC RS flip-flop) is shown constructed of NOR gates.

The flip-flop is so configured that the output of each gate is connected to an input of the other. The Q and Q' outputs are called the *principal* and *complement* outputs of the flip-flop, respectively. A little thought will reveal that, in the absence of input signals S and R, there are two unconditionally stable states that the flip-flop may assume (hence the term bistable). These are $Q = 1, Q' = 0$ and $Q = 0, Q' = 1$. That is, a logical-1 output on Q will be inverted by the first gate to produce a logical-0 output on Q' and vice versa.

The "state" of the flip-flop is understood to mean the logical value of the Q output. One may consider that either a 0 or a 1 is stored in the flip-flop. The state of the flip-flop is changed by signals at the inputs S (set) and R (reset). Reference to the equations for a NOR gate will confirm that the following truth table applies to the flip-flop:

R	S	Q	Q'
0	0	Q	$\bar{Q}$
0	1	1	0
1	0	0	1
1	1	0	0*

*For period in which both R and S are 1. After R and S go to 0, the state will be determined by the order of removal of the inputs.

Clocked Flip-flops. For many applications it is convenient to isolate the inputs from the flip-flop except for brief sampling periods. This opera-

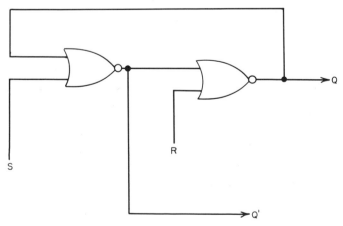

Fig. 7-1   DC RS flip-flop.

tion is accomplished by gating the inputs with a sampling pulse or "clock pulse," which initiates the action of the flip-flop. If periodic clock pulses are employed in a system, then all actions of the system are timed to the discrete intervals of the clock-pulse period. Our attention may thus be focused on the time increments at which each successive clock pulse appears, rather than on the continuum of times at and within the increments.

A variety of clocked flip-flops with useful characteristics might be proposed. We introduce here types designated as RS, JK, T, and D.* A sample circuit for the clocked RS flip-flop is shown in Fig. 7-2. The configuration is identical to the DC RS flip-flop, except that two additional inputs S_C and R_C are gated to the flip-flop via AND gates enabled by the clock pulse C_P. The direct inputs, now labeled R_D and S_D, may also be present for use in presetting the flip-flop to a desired initial state, independent of the clock pulse. With no electrical connection to the direct inputs, changes in the state of the flip-flop can occur only during those periods of time when the clock pulse is present. It is desirable to limit the number of state changes of the flip-flop to one per clock pulse. The circuit of Fig. 7-2 is not particularly promising from this standpoint, since changes can continue to occur as long as the clock pulse is present. Should the inputs R_C and S_C be derived from the outputs Q and Q', for example, a clock pulse of duration greater than the switching time of the flip-flop could give rise

*The flip-flop terminology follows that of Phister (see Ref. 3).

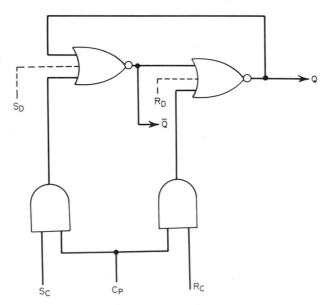

Fig. 7-2 Clocked RS flip-flop.

to a series of changes, and leave the final state of the flip-flop ambiguously defined. Additional circuitry required for reliable triggering is discussed in a later section. For the time being we will assume that sufficiently narrow clock pulses are used to assure that one change only will occur during any clock period.

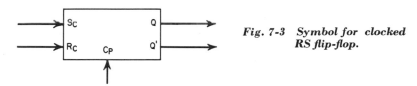

Fig. 7-3 **Symbol for clocked RS flip-flop.**

The discrete periods in which no changes can occur in the flip-flop state are termed clock intervals and may be labeled successively $1, 2, 3, \ldots, t$, $t + 1, \ldots$, etc. Thus, if the present flip-flop state is Q_t, the next state (after the next clock pulse) is labeled Q_{t+1}. With this in mind, a truth table for a clocked RS flip-flop may be developed as shown below. The Q' output is the complement of the Q output ($Q' = \overline{Q}$). This fact will be assumed for all clocked flip-flops unless otherwise stated.

R_C	S_C	Q_{t+1}
0	0	Q_t
0	1	1
1	0	0
1	1	?

The last-row entry results in an arbitrary state of the flip-flop, and is normally avoided.

Flip-flops of the type just discussed are normally represented by rectangular symbols, as shown in Fig. 7-3. Most clocked flip-flops will also have at least one direct-coupled R_D or S_D input.

A truth table for the JK flip-flop follows:

J	K	Q_{t+1}
0	0	Q_t
0	1	0
1	0	1
1	1	$\overline{Q}_t$

The last-row entry indeterminancy which appeared in the RS truth table is absent from this flip-flop. When both J and K are 1, the flip-flop will reliably change state.

The T flip-flop has a single input with characteristics as follows:

T	Q_{t+1}
0	Q_t
1	$\bar{Q}_t$

In the presence of a 1 input on T, the flip-flop will change state with each clock pulse.

The characteristics of a D flip-flop are shown below:

D	Q_{t+1}
0	0
1	1

The output at time $t + 1$ will assume the state of the input at time t.

Design Examples

The essence of sequential design is to formulate logical equations for the inputs of the flip-flops in terms of system parameters, which include input parameters to the system and the present flip-flop outputs. The specific steps employed are best introduced by means of examples. In this section we consider some common configurations. The section following this one then treats the general approach to sequential design.

Binary Counter. A binary counter is a series of flip-flops whose states correspond to a binary number. This number is increased by one every time a clock pulse occurs. Suppose we want a counter with a capacity of eight (binary 0 to 7). The desired sequence is:

0	0	0
0	0	1
0	1	0
0	1	1
1	0	0
1	0	1
1	1	0
1	1	1
0	0	0

etc.

Every eighth count, the counter recycles to 000. We identify each significant digit with the output of a flip-flop. Three flip-flops are therefore required; we will employ the RS flip-flop types. Design information may be listed in a table, as shown in Table 7-1. The subscripted variables

TABLE 7-1 Design Table for Binary Counter

Present state (time t)			Next state (time $t+1$)			Required flip-flop inputs					
Q_3	Q_2	Q_1	Q_3	Q_2	Q_1	R_3	S_3	R_2	S_2	R_1	S_1
0	0	0	0	0	1	×	0	×	0	0	1
0	0	1	0	1	0	×	0	0	1	1	0
0	1	0	0	1	1	×	0	0	×	0	1
0	1	1	1	0	0	0	1	1	0	1	0
1	0	0	1	0	1	0	×	×	0	0	1
1	0	1	1	1	0	0	×	0	1	1	0
1	1	0	1	1	1	0	×	0	×	0	1
1	1	1	0	0	0	1	0	1	0	1	0

R_3, S_3, R_2, S_2, R_1, S_1 are the R_C and S_C inputs to the flip-flops whose principal outputs are correspondingly subscripted.

Let us consider how the required inputs are determined. We start with the present state of the flip-flops at some time t. The inputs to these flip-flops at time t should be of such value as to advance the counter one count upon receipt of a clock pulse.

The inputs R_3, S_3, R_2, S_2, R_1, and S_1 will thus be functions of the present flip-flop states. By observing the required change in a flip-flop at the next clock pulse, appropriate values of the inputs of that flip-flop may be listed. A × indicates a "don't care."

As an example, consider the entries in the first row of Table 7-1. The flip-flops are in state $Q_3Q_2Q_1 = 000$. The next state is $Q_3Q_2Q_1 = 001$. Flip-flop Q_3 remains at 0 after the clock pulse. We certainly want $S_3 = 0$ at time t, because $S_3 = 1$ would, at time $t + 1$, either set $Q_3 = 1$ or cause an indeterminant state (see truth table for RS flip-flop). What value should R_3 have? We examine both possibilities. For $S_3 = 0$ and $R_3 = 1$, Q_3 will be at 0 at time $t + 1$. For $S_3 = 0$ and $R_3 = 0$, Q_3 will be the same at time $t + 1$ as it was at time t. Either value of R_3 will result in $Q_3 = 0$ at time $t + 1$. The required value of R_3 is therefore "don't care" or ×. The required values of R_2 and S_2 are found in an identical manner.

Flip-flop Q_1 changes from 0 to 1. This transition requires $S_1 = 1$ and $R_1 = 0$. The remaining rows may be determined in a similar fashion. The following transition rules are helpful:

Q (time t)	Q (time $t+1$)	R_C (time t)	S_C (time t)
0	0	$\times$	0
0	1	0	1
1	0	1	0
1	1	0	$\times$

It now remains to determine the equations for the R and S inputs in terms of the present-state Q's. Six Karnaugh maps for these variables are drawn in Fig. 7-4.

We obtain

$$R_3 = Q_3 Q_2 Q_1$$
$$S_3 = \bar{Q}_3 Q_2 Q_1$$
$$R_2 = Q_2 Q_1$$
$$S_2 = \bar{Q}_2 Q_1 \qquad\qquad (7\text{-}1)$$
$$R_1 = Q_1$$
$$S_1 = \bar{Q}_1$$

A block diagram is drawn in Fig. 7-5.

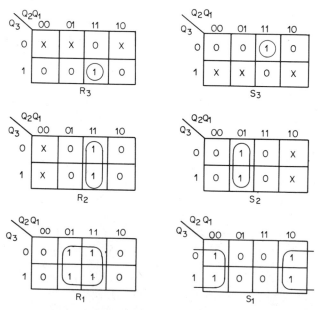

Fig. 7-4   Karnaugh maps for three-stage binary counter implemented with RS flip-flops.

Decade Counter. A counter which recycles after n distinct counts is called a mod n counter. The three-stage binary counter of the previous example is thus a mod 8 counter. We consider a counter which runs through ten distinct states, therefore, a decade counter. Decade counters have considerable application because our decimal system of counting is mod 10.

The minimum number of flip-flops required to construct a mod n counter is the integer m which satisfies the relation

$$m \geq \log_2 n > m - 1 \tag{7-2}$$

For a decade counter, $\log_2 10 = 3.3$. We therefore need four flip-flops. The actual states through which we cycle the counter need not correspond to the binary values of the counts since we have complete freedom to interpret the sets of flip-flop outputs. Often, practical considerations (such as minimization of gates required) govern the sequence of states through which we decide to cycle the counter.

An important practical consideration is the error-free interpretation of each count for purposes of sequencing operations. Ideally, we would like each transition between states to be abrupt and precise, i.e., for each flip-flop to switch at exactly the same time. Realistically, differences in wire path lengths and circuit parameters introduce uncertainties during the brief switching period. For example, consider the transition between the states $Q_3Q_2Q_1 = 011$ and $Q_3Q_2Q_1 = 100$ in the binary counter of the previous example. This transition requires all three flip-flops to changes states. If Q_3 responds faster to the clock-pulse trigger than Q_2, and Q_2 faster than Q_1, the counter would progress through the transient states 111 and 101 before reaching the desired 100 state. Thus, in going from a binary 3 to a binary 4 count, we might also initiate actions corresponding to a binary 7

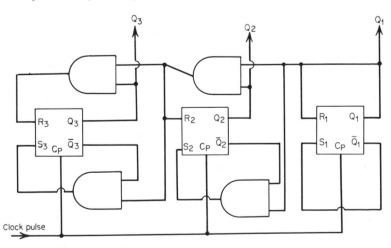

Fig. 7-5 Three-stage binary counter.

and a binary 5. If the switching speeds of the flip-flops are arbitrary, we can, in fact, go through any binary state during transition. To prevent misinterpretation, the following steps may be taken:

1. Use of a two-phase clock. Two clock pulses are used during each clock increment. One pulse advances counters and initiates other flip-flop actions. A second pulse, delayed in time with respect to the first, "strobes" or enables the interpretation of the states after the transient periods.

2. Desensitization of the interpretation circuitry. The interpretative circuitry is "slowed down" to be insensitive to short transients.

3. Use of a single-transition count sequence. The counters are designed to sequence so that the transition from any state to the succeeding one involves a change in the logical value of only one flip-flop.

The decade counter will be designed to satisfy a single-transition count sequence of step 3 above. We arbitrarily choose the following such sequence: $Q_4Q_3Q_2Q_1 = 0000, 0001, 0011, 0010, 0110, 0111, 0101, 1101, 1100, 0100, 0000, \ldots$.

Any of the four flip-flop types introduced in the chapter may be used for the design. The transition rules governing the input requirements for RS, JK, T, and D flip-flop types are listed in Table 7-2. We will employ

TABLE 7-2   Flip-flop Transition Input Requirements

Q (time t)	Q (time $t+1$)	R_C	S_C	J_C	K_C	T_C	D_C
0 $\longrightarrow$ 0		$\times$	0	0	$\times$	0	0
0 $\longrightarrow$ 1		0	1	1	$\times$	1	1
1 $\longrightarrow$ 0		1	0	$\times$	1	1	0
1 $\longrightarrow$ 1		0	$\times$	$\times$	0	0	1

JK flip-flops for the illustration. Table 7-3 shows the design table for JK flip-flops. The design using the other types is left as an exercise (Probs. 7-1 and 7-2).

The input requirements are plotted in Fig. 7-6. We obtain the problem solution

$$J_4 = Q_3\bar{Q}_2Q_1$$
$$K_4 = \bar{Q}_1$$
$$J_3 = Q_2\bar{Q}_1$$
$$K_3 = \bar{Q}_4\bar{Q}_2\bar{Q}_1$$
$$J_2 = \bar{Q}_3Q_1$$
$$K_2 = Q_3Q_1$$
$$J_1 = \bar{Q}_2 + Q_3$$
$$K_1 = \bar{Q}_3Q_2 + Q_4Q_1$$

(7-3)

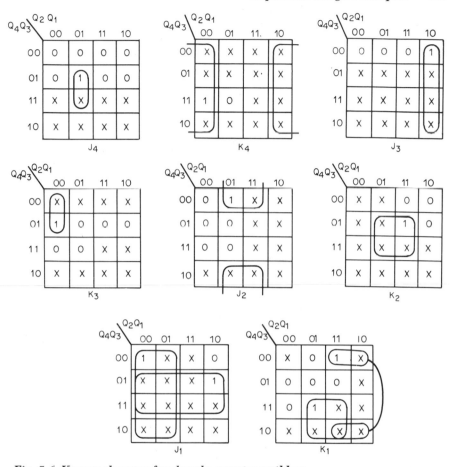

Fig. 7-6 Karnaugh maps for decade-counter problem.

TABLE 7-3 Input Requirements for *JK* Flip-flops

Present states				Next states				Required inputs							
Q_4	Q_3	Q_2	Q_1	Q_4	Q_3	Q_2	Q_1	J_4	K_4	J_3	K_3	J_2	K_2	J_1	K_1
0	0	0	0	0	0	0	1	0	$\times$	0	$\times$	0	$\times$	1	$\times$
0	0	0	1	0	0	1	1	0	$\times$	0	$\times$	1	$\times$	$\times$	0
0	0	1	1	0	0	1	0	0	$\times$	0	$\times$	$\times$	0	$\times$	1
0	0	1	0	0	1	1	0	0	$\times$	1	$\times$	$\times$	0	0	$\times$
0	1	1	0	0	1	1	1	0	$\times$	$\times$	0	$\times$	0	1	$\times$
0	1	1	1	0	1	0	1	0	$\times$	$\times$	0	$\times$	1	$\times$	0
0	1	0	1	1	1	0	1	1	$\times$	$\times$	0	0	$\times$	$\times$	0
1	1	0	1	1	1	0	0	$\times$	0	$\times$	0	0	$\times$	$\times$	1
1	1	0	0	0	1	0	0	$\times$	1	$\times$	0	0	$\times$	$\times$	0
0	1	0	0	0	0	0	0	0	$\times$	$\times$	1	0	$\times$	$\times$	0
—	—	—	—												
0	0	0	0												

Shift Register. A shift register is a subsystem composed of n series-connected storage elements (flip-flops) which transfer the contents of each element to its respective succeeding element upon receipt of a clock pulse. A block diagram of a shift register is shown in Fig. 7-7.

Let us examine the particular case of a five-stage shift register. The register is initially set to zero. The following time-dependent action occurs, with an assumed input sequence of 1101101 . . . , etc.:

Time	Input	Q_1	Q_2	Q_3	Q_4	Q_5
0	1	0	0	0	0	0
1	1	1	0	0	0	0
2	0	1	1	0	0	0
3	1	0	1	1	0	0
4	1	1	0	1	1	0
5	0	1	1	0	1	1
6	1	0	1	1	0	1
etc.	etc.	1	0	1	1	0

The shift register is useful for storing and processing sets of data in serial fashion. Sample applications include
1. Arithmetic operations (adding, multiplying)
2. Addressing
3. Circulating storage
4. Sequence generation (counter)

We will consider two specific cases: an addition problem and a counter. The reader is referred to the cited chapter references for examples of other applications.

Serial Adder. The typical serial addition problem involves two registers: an accumulator register which stores the addend, and a second register which stores the augend. The sum of two numbers is routed back to the accumulator register, replacing the addend. The accumulator must therefore have sufficient capacity to store the sum of the two numbers.

A block diagram of a serial-adder configuration is shown in Fig. 7-8. The accumulator and augend registers are shown as six-cell registers. The block identified as "sum" performs the full-sum function discussed in

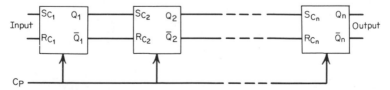

Fig. 7-7   Shift register.

connection with Eq. (6-20). The flip-flop labeled D delays the carry from the previous addition to be added into the next most significant addition.

Starting with the least significant bits, the contents of the two registers are shifted to the right with each clock pulse. The sum is formed and returned to the accumulator. The total addition requires six clock-pulse intervals.

Sequence Generator. The introduction of feedback from flip-flops in a shift-register chain to the input can provide a convenient means of generating a sequence of states. One of the simplest configurations is a Johnson counter, formed by cross-connecting the outputs of the last stage of a shift register to the inputs of the first state. A ten-state Johnson counter using five RS flip-flops is shown in Fig. 7-9. Assuming an initial state of 00000, the ten states generated are:

0	0	0	0	0
1	0	0	0	0
1	1	0	0	0
1	1	1	0	0
1	1	1	1	0
1	1	1	1	1
0	1	1	1	1
0	0	1	1	1
0	0	0	1	1
0	0	0	0	1
—	—	—	—	—
0	0	0	0	0

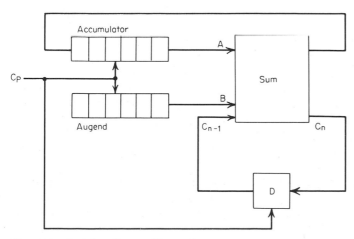

Fig. 7-8 Serial-adder configuration.

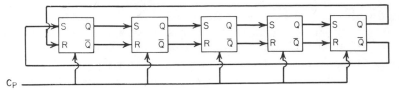

Fig. 7-9   Ten-state Johnson counter.

The Johnson counter has the attractive feature of only one change in flip-flop state per clock pulse, design simplicity, and no external gating. A disadvantage is the extra flip-flop required over a gated mod 10 counter.

Sequential-design Procedure*

The simple circuits considered thus far, and others like them, may be combined to perform an endless variety of processing functions. The designs illustrate the translation of subsystem requirements into functional hardware. Often requirements are not as succinctly stated as those for a counter or other simple circuits. A good example is that of a sequence detector which might have the specification, *Design a circuit with an input which receives single-digit binary numbers synchronized with the clock-pulse generator. The output of the circuit, initially set to logical 0, will go to logical 1 if three or more consecutive logical 1s appear in the input line. It will remain in the logical 1 state until two consecutive logical 0s appear on the input line.* For such a problem, we clearly need a general approach to the design of sequential machines. Although the topic is a complex one, we shall briefly sketch the basic procedure for such design problems.

The interested reader may pursue the topic further in Refs. 1 (intermediate level) or 2 (advanced level). The procedure may be conveniently divided into three tasks: problem specification, assignment of states, and formulation of flip-flop and output equations. Problem specification is the reduction of the system description into well-defined memory states, and the delineation of a set of rules for transitions between these states. The assignment of states involves equating the defined memory states with the logical values of one or more flip-flops. The equations for the inputs to the flip-flops and the outputs of the system are then derived from the transition rules and state assignments.

The procedure will be illustrated by carrying through a design of the sequence detector just specified.

Problem Specification.   We define a machine by a series of distinct states (or memory states). The transition from a given state to a new state is determined by the given state and the input to the system. For the

*This section may be omitted on first reading.

sake of generality, the memory states at this point will be designated by letters of the alphabet. For each state we assign a value to the output variable.* A block diagram of the machine is drawn in Fig. 7-10. We have designated the input variable as X and the output variable as Z.

Assume the machine is initially in memory state a. We assign a 0 output to a as required by the problem specification. Suppose the first input is a 0. There is no reason for the machine to leave the initial state because its future behavior would not in any way be altered by an initial 0 input. If the first input is a 1, however, the machine must record this event by going to memory state b. The output remains at 0. State b signifies that the first of a possible sequence of three consecutive 1s has been received. If 1s continue to be received at succeeding clock intervals, the machine will progress through state c (two consecutive 1s) to state d (three or more consecutive 1s). We assign an output of 1 to state d. If an input of 0 is received prior to arriving at state d, thus breaking the sequence of 1s, the machine will return to the initial state to begin again to look for three consecutive 1s. Assuming now that the machine has reached state d, it will then look for two consecutive 0s. At the first 0 it will progress to state e, with an output still equal to 1. If another 0 follows, it will return to state a, but if a 1 occurs, it will step back to state d. The memory states, the output of the states, and transition rules between the states are thus defined, and the reader may verify that they meet the specifications of the sequence detector.

An abbreviated listing of the above discussion is contained in the state table of Table 7-4. The table identifies the progression between the present states and the next states according to the inputs received, and the assignment of output values to the present states. For example, the first row indicates that if the machine is in state a, it will remain there with a 0 input but progress to state b with a 1 input. It shows that state a is assigned an input value of 0.

Although this state table has been developed for a specific problem, its general construction should be clear. A memory state and its associated

---

*Such a machine is called a Moore machine. Alternatively, a Mealey machine is one whose output depends upon both the internal state of the machine and the present value of the input.

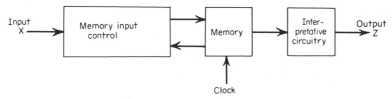

Fig. 7-10 Block diagram of sequence detector.

TABLE 7-4   State Table for Sequence-detector Problem

Present state	Next state $X = 0$	$X = 1$	Output of present state Z
a	a	b	0
b	a	c	0
c	a	d	0
d	e	d	1
e	a	d	1

outputs are defined for each unique situation the machine must record. The rules for transitions between the states, as a function of the input(s) and the present state, are indicated. If two states have the same output, and the future behavior of the machine is identical from either state, then the two states are equivalent.

The three-stage binary counter designed previously in this chapter may be similarly tabulated. It has no input and three outputs. A state table for this machine is shown in Table 7-5.

TABLE 7-5   State Table for Three-stage Binary Counter

Present state	Next state	Outputs Z_3	Z_2	Z_1
a	b	0	0	0
b	c	0	0	1
c	d	0	1	0
d	e	0	1	1
e	f	1	0	0
f	g	1	0	1
g	h	1	1	0
h	a	1	1	1

State Assignment. The second task in the sequential-design procedure is to relate the memory states a, b, c, ... to the logical values of storage elements. The sequence-detector problem has five memory states; therefore, by Eq. (7-2), at least three flip-flops are required. To each memory state is assigned a unique set of flip-flop logical-output values. The specific assignment will determine the combinatorial circuitry required to realize the appropriate flip-flop-input functions and the machine-output function. Methods exist for achieving a "good" assignment which require a relatively small amount of combinatorial circuitry (Ref. 2). On the other hand, we might wish to minimize transient hazards, as discussed previously. We will select the following assignment:*

Memory state	Flip-flop states Q_3		Q_1
	Q_3	Q_2	Q_1
a	0	0	0
b	0	0	1
c	0	1	1
d	1	1	1
e	1	1	0

If we substitute the above assignment into Table 7-4, we obtain in Table 7-6 the assigned-state table for the sequence detector.

Formulation of Equations. Having proceeded to Table 7-6, the die is cast, and there remains only the job of writing equations for the flip-flop inputs and the machine output.

TABLE 7-6 **State Table for Sequence Detector after Assignment**

Present states Q_3		Q_1	Next states $X = 0$			$X = 1$			Output Z
Q_3	Q_2	Q_1							
0	0	0	0	0	0	0	0	1	0
0	0	1	0	0	0	0	1	1	0
0	1	1	0	0	0	1	1	1	0
1	1	1	1	1	0	1	1	1	1
1	1	0	0	0	0	1	1	1	1

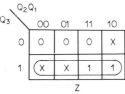

Fig. 7-11 Plot of sequence detector output.

First, consider the output. It is plotted as a function of the present states in the Karnaugh map of Fig. 7-11. We have

$$Z = Q_3$$

The fact that no gates are required for the interpretative circuitry was anticipated by the state assignment.

Next we consider the equations for the flip-flop inputs. We assume JK flip-flops. The next states are functions not only of the present states but also of the input. Table 7-7 is the familiar design

*We choose the state of Q_3 to coincide with the output requirements. This will simplify the interpretative circuitry.

TABLE 7-7   Design Table for Sequence Detector

Input and present states				Next states			Flip-flop input requirements					
X	Q_3	Q_2	Q_1	Q_3	Q_2	Q_1	J_3	K_3	J_2	K_2	J_1	K_1
0	0	0	0	0	0	0	0	×	0	×	0	×
1	0	0	0	0	0	1	0	×	0	×	1	×
0	0	0	1	0	0	0	0	×	0	×	×	1
1	0	0	1	0	1	1	0	×	1	×	×	0
0	0	1	1	0	0	0	0	×	×	1	×	1
1	0	1	1	1	1	1	1	×	×	0	×	0
0	1	1	1	1	1	0	×	0	×	0	×	1
1	1	1	1	1	1	1	×	0	×	0	×	0
0	1	1	0	0	0	0	×	1	×	1	0	×
1	1	1	0	1	1	1	×	0	×	0	1	×

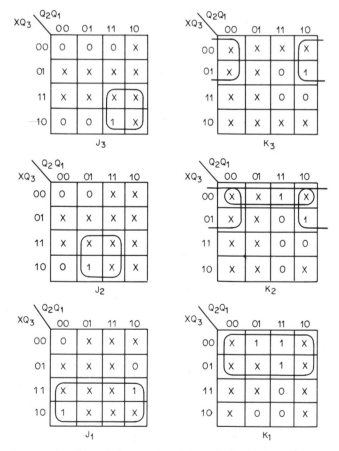

Fig. 7-12   Karnaugh maps of input functions for sequence-detector problem.

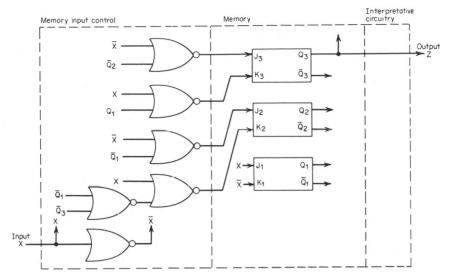

Fig. 7-13 **Completed design of sequence detector.**

table formed by rearranging Table 7-6. We fill in the input requirements in the normal manner and plot these functions in Fig. 7-12. The input equations are

$$J_3 = XQ_2$$
$$K_3 = \bar{X}\bar{Q}_1$$
$$J_2 = XQ_1$$
$$K_2 = \bar{X}\bar{Q}_1 + \bar{X}\bar{Q}_3$$
$$J_1 = X$$
$$K_1 = \bar{X}$$

A block diagram of the design, implemented with NOR gates, is shown in Fig. 7-13.

Reliable Triggering Methods*

Reliable triggering of clocked flip-flops is essential to the design of digital machines. As mentioned previously, assurance must be obtained that only one state-change of a clocked flip-flop will occur coincident with the clock pulse. Such assurance is not obtained with the circuit of Fig. 7-2. The circuit will continue to change state as long as the clock pulse is present. For example, if this flip-flop were employed as the first stage of the binary counter of Eq. (7-1), we would have $R = Q$ and $S = \bar{Q}$. Suppose that, prior to some clock pulse, $Q = 0$. At the incidence of the clock pulse, Q would be set to 1, R would be 1, and the flip-flop would attempt to reset Q

*This section may be omitted on first reading.

back to 0. If the clock pulse is longer in duration than the flip-flop switching speed, the flip-flop would oscillate between the set and reset conditions and leave the final outcome in doubt.

A variety of methods may be employed to obtain reliable triggering action. An obvious method is to use very narrow clock pulses. This solution, however, immediately runs into difficulties. The flip-flop switching speed can be on the order of nanoseconds. The clock pulse must be long enough to cause the flip-flop transistors to reach the switching point, but sufficiently short to prevent two switching actions from occurring. Variabilities in switching speed of flip-flops from device to device result in different optimum clock-pulse width. A situation could conceivably arise where no single clock-pulse width would satisfactorily operate all flip-flops in a system. Moreover, this sort of "tinkering" with basically solid-acting digital circuits is unsatisfactory from a systems-reliability standpoint.

The two most prevalent means for achieving reliable triggering action are through the capacitance-coupled and the master-slave configurations. The former is sensitive to clock-pulse transitions, the latter to clock-pulse levels. We now discuss the two methods.

Capacitance-coupled Configuration.   The capacitance-coupled configuration employs a basic differentiating circuit. The R and S inputs are

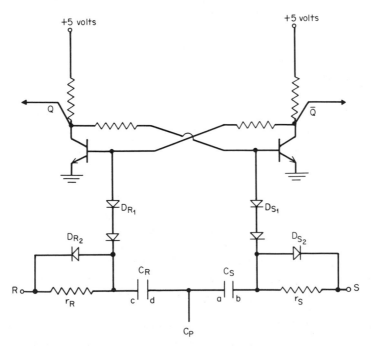

Fig. 7-14   Capacitance-coupled clock arrangement.

stored on separate capacitors and released to the flip-flop by action of the trailing edge of the clock pulse. Since the state change occurs at the end of the clock pulse, only one flip-flop state change is possible.

A sample circuit is shown in Fig. 7-14. The cross-coupled transistors form a basic DC RS flip-flop with outputs Q and $\bar{Q}$. The flip-flop communicates with the R and S inputs via the paths through D_{S_1} or D_{R_1}. To visualize the circuit operation, assume that it is desired to set $Q = 1$. S is at $+5$ volts (logical 1), R is at ground (logical 0), and the clock input C_P is initially at ground. D_{S_1} and D_{R_1} are nonconducting since the offset voltage of the serial diode pairs is greater than the base voltage of the transistors. Capacitor C_S is charged to $+5$ volts via the S input and resistor r_S. Points a, c, and d are at ground and point b is at $+5$ volts. Now assume C_P rises to $+5$ volts. C_S immediately discharges through D_{S_2}, and C_R charges through D_{R_2}. Points a, b, and d are now at $+5$ volts, and point c is at ground. When C_P falls back to ground, capacitor C_R cannot immediately discharge since D_{R_2} is in a reverse-biased direction for the discharge current. The discharge rate is consequently limited by the $r_R C_R$ time constant. With point b at ground, point c is forced to go to a negative voltage to maintain the charged state of C_R. Diodes D_{R_1} conduct, and the transistor associated with the Q output is turned off. The Q output voltage thus rises to a logical-1 voltage. Diodes D_{S_1} do not conduct, on the other hand, since point b is not depressed below ground potential. Reversing the actions of the R and S circuit segments, a similar description applies for resetting $Q = 0$. A truth table for the flip-flop is shown in Table 7-8. The characteristics look somewhat different from those previously discussed, but coincide with those of an RS flip-flop if we define logical-0 and logical-1 levels as $+5$ and 0 volts, respectively, and exchange the labeling of the Q and $\bar{Q}$ outputs.

TABLE 7-8   Truth Table for Capacitance-coupled Flip-flop

R	S	Q_{t+1}
0	0	?
0	1	1
1	0	0
1	1	Q_t

The capacitance-coupled input is sometimes called a pedestal gate.

Master-Slave Configuration.   The master-slave configuration is an asynchronous design employing two (or more) DC RS flip-flops. The two DC flip-flops combine to behave like a single synchronous or clocked flip-flop. The clock-pulse input is treated as a level input in tne same manner as the R and S (or J and K) inputs. One of the flip-flops, the "master," is a control circuit which determines the state of the second flip-flop, the "slave." The outputs are taken from the slave.

The design of the master-slave configuration may be implemented using the sequential-design procedure previously outlined in the chapter. In using the procedure for synchronous design, however, one must keep in mind that the clock pulse is an implicit input to the machine and state transitions do not inherently occur at clock-pulse times but whenever the input levels change. In other words, the synchronous nature of the machine must be explicitly designed into it. Since these procedural modifications for asynchronous design are generally useful, we will illustrate a design of the master-slave configuration in this section. The following characteristics are assumed:

1. There are three inputs to the system, labeled R, S, and C (the clock pulse).

2. R and S can never simultaneously assume the logical value of 1.

3. When the clock pulse C occurs ($C = 1$), changes in R and S are forbidden. This condition prevents ambiguities in the outcome, and is a reasonable constraint for narrow clock pulses.

Table 7-9 is a state table for the master-slave configuration. We assume that the machine is initially in state a with an output of 0. The machine

TABLE 7-9 State Table for Master-Slave Configuration

Present state	Next state after input *RSC*						Output Q
	000	001	010	011	100	101	
a	a	a	a	b	a	a	0
b	. . .	. . .	c	b	. . .	. . .	0
c	c	c	c	c	c	d	1
d	. . .	. . .	. . .	. . .	a	d	1

remains in state a until $R = 0$, $S = 1$, and $C = 1$. The machine then goes into the "holding" state b, still with an output of 0. The function of state b is to recognize that a set condition has occurred, but to hold off a change in the output until the clock pulse returns to 0. If we allowed the output to go immediately to 1, then the R and S inputs might subsequently change value and the machine would resemble the unsatisfactory design of Fig. 7-2.

When in state b, the only other input condition that can occur is $R = 0$, $S = 1$, and $C = 0$. This follows from statement 3 above, and indicates that the clock pulse has returned to zero. The forbidden-input states may be used as "don't care" conditions. With the clock pulse back at zero, the machine enters state c with an output of 1. It will stay in state c until the reset condition $R = 1$, $S = 0$, $C = 1$ occurs. The machine then goes into holding state d to await the end of the clock pulse, and subsequently back to state a.

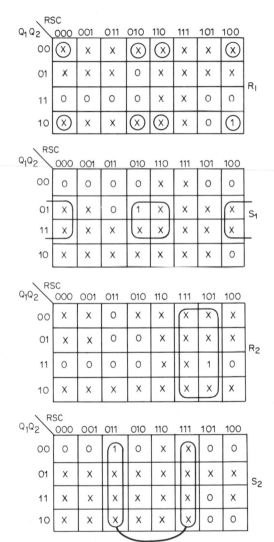

Fig. 7-15   Karnaugh maps for input functions of master-slave configurations.

In Table 7-10 is listed the state table with the following assignments made:

$$a = 00$$
$$b = 01$$
$$c = 11$$
$$d = 10$$

This assignment not only has the advantage of a single flip-flop change between transitions, but also allows the output of Q_1 to be directly used as the machine output.

The next step is to derive a design table so that the input requirements for R_1, S_1, R_2, and S_2 may be delineated. The reader is urged to follow

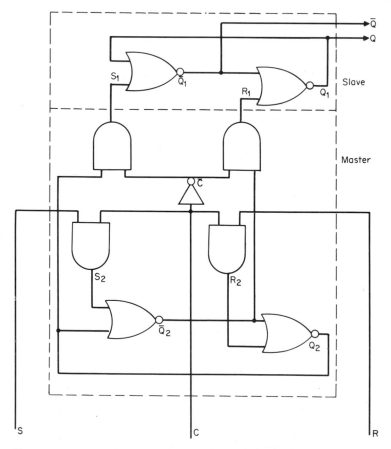

Fig. 7-16 *Master-slave synchronous flip-flop design.*

TABLE 7-10   State Table for Master-Slave Configuration after State Assignment

Present states Q_1Q_2	Next states after input RSC						Output Q
	000	001	010	011	100	101	
00	00	00	00	01	00	00	0
01	...	...	11	01	...	...	0
11	11	11	11	11	11	10	1
10	...	...	...	...	00	10	1

through with this step and demonstrate the validity of the Karnaugh maps of Fig. 7-15. The result is

$$R_1 = \bar{Q}_2 C$$
$$S_1 = Q_2 C$$
$$R_2 = RC$$
$$S_2 = SC$$

and the design is shown in Fig. 7-16.

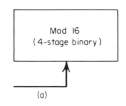

(a)

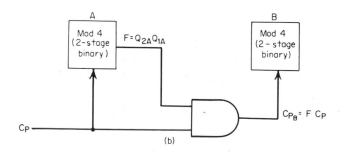

(b)

Fig. 7-17   Alternative methods. (a) Single counter. (b) Series-connected counter for achieving mod 16 counter.

Series-connected Design

As we have seen, the basic method for constructing counters is to determine the inputs required to achieve the next state for each flip-flop in terms of the present states of all of the flip-flops. It is often desirable to employ an alternative procedure to the construction of the counters: that of clock-pulse gating. The concept behind this method is to use two or more smaller counters in series, in lieu of one large counter.

As an example, suppose we wish to count up to 15 events. Two ways of achieving a mod 16 counter are indicated in Fig. 7-17.

In Fig. 7-17a, the mod 16 counter is a four–flip-flop system with the following gating required:

$$R_4 = Q_4 Q_3 Q_2 Q_1$$
$$S_4 = \bar{Q}_4 Q_3 Q_2 Q_1$$
$$R_3 = Q_3 Q_2 Q_1$$
$$S_3 = \bar{Q}_3 Q_2 Q_1$$
$$R_2 = Q_2 Q_1$$
$$S_2 = \bar{Q}_2 Q_1$$
$$R_1 = Q_1$$
$$S_1 = \bar{Q}_1$$

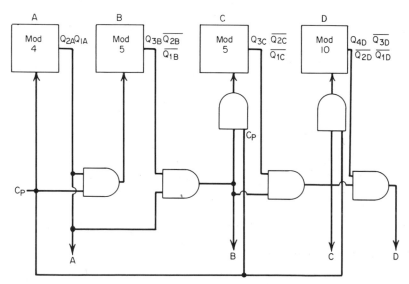

Fig. 7-18 Mod 1,000 counter with periodic action requirements.

In Fig. 7-17b, the mod 16 counter is composed of two mod 4 counters in such a manner that counter B is clocked only when counter A is in the fourth state (binary 3). The appropriate gating equations are

$$R_{2A} = Q_{2A}Q_{1A}$$
$$S_{2A} = \bar{Q}_{2A}Q_{1A}$$
$$R_{1A} = Q_{1A}$$
$$S_{1A} = \bar{Q}_{1A}$$
$$R_{2B} = Q_{2B}Q_{1B}$$
$$S_{2B} = \bar{Q}_{2B}Q_{1B}$$
$$R_{1B} = Q_{1B}$$
$$S_{1B} = \bar{Q}_{1B}$$
$$C_{P(B)} = Q_{2A}Q_{1A}C_{P}$$

A clock-pulse-gated counter is especially useful when periodic actions must be initiated during a long count. For example, suppose in a 1,000-step system action A is to occur every fourth count, action B every twentieth count, action C every hundredth count, and action D every thousandth count. The use of a 10-stage mod 1,000 counter would be prohibitive in gating requirements, from the standpoint of both counter equations and selection network. A practical design is shown in Fig. 7-18.

A special and important case of series-connected design employs only one flip-flop per serial counter (mod 2). The output of one counter is used as the clock input to the next. The first counter is operated by an external clock pulse. This often-used design, called a ripple-through counter, requires no input gating for binary counts (counts expressible as 2^n, where n is the number of serial states). A three-stage binary ripple-through counter is shown in Fig. 7-19.

The advantage of ripple-through counters is their extreme simplicity. A disadvantage is the settling time required for all changes to propagate down (or "ripple") through the counter starting from the least significant stage. During the settling time, transient states occur, but these states are predictable and their effects can be accommodated in design.

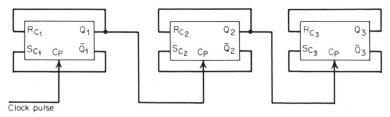

Fig. 7-19   Three-stage ripple-through counter.

REFERENCES

1. Bartee, Thomas C., Irwin L. Lebow, and Irving S. Reed: *Theory and Design of Digital Machines*, McGraw-Hill, 1962.
2. Hartmanis, J., and R. E. Stearns: *Algebraic Structure Theory of Sequential Machines*, Prentice-Hall, 1966.
3. Phister, Montgomery, Jr.: *Logical Design of Digital Computers*, John Wiley, 1958.

PROBLEMS

7-1. Design a four-stage decade counter using *RS* flip-flops. The desired count sequence is:

Time	Q_4	Q_3	Q_2	Q_1
t	0	0	0	0
$t+1$	0	0	0	1
$t+2$	0	0	1	1
$t+3$	0	0	1	0
$t+4$	0	1	1	0
$t+5$	0	1	1	1
$t+6$	0	1	0	1
$t+7$	1	1	0	1
$t+8$	1	1	0	0
$t+9$	0	1	0	0
$t+10$	0	0	0	0
etc.				

7-2. Repeat Prob. 7-1 using first *T* flip-flops, then *D* flip-flops.

7-3. Design a synchronous sequence detector with a single input X and a single output Z. The output, initially at 0, is required to go to 1 whenever the input sequence is 110010. The output then returns to 0 after two clock intervals.

7-4. In the processing of a razor blade from the drawn metal, seven steps are required in sequence:
  1. Cut.
  2. Rough sharpen.
  3. Fine sharpen.
  4. Rough strop.
  5. Fine strop.
  6. Print trademark.
  7. Store in shute.
  Every fifth blade, however, requires an additional step:
  8. Package this and preceding four blades in dispenser.
Design a digital control system for the process. The control system provides signals that initiate each of the actions of the processor. The clock rate is adjusted to permit sufficient time for processing between clock pulses. Use *JK* flip-flops for clocked counter needs.

EIGHT

Characteristics of Digital Integrated Circuits

The implementation of a digital machine requires two design activities: first the logical design, and second the physical realization of that design. Very often, the dictates of the logical design must be modified by a practical consideration of the allowed interconnections between circuits, limitations on the magnitudes of currents and voltages, and nonnegligible times required for the propagation of electrical signals through the circuits.

The design of digital systems with commercial integrated circuits has been made convenient by manufacturers who provide simple design rules, optimum configurations, and other literature on applications. Nevertheless, a sensible selection of integrated-circuit type, the combined use of integrated circuits from two or more vendors, and the design of peripheral equipment to interface with the digital system require an understanding of the electrical properties of integrated circuits and an ability to distinguish the pros and cons of the different available types. A start toward this end is the purpose of this chapter.

Basic Definitions

Manufacturers generally supply the variety of gates, flip-flops, and sub-assemblies required for a complete digital design as members of a digital "family" with compatible characteristics. The fundamental building block of the family is the simple inverting gate (either a NAND or a NOR gate), and the more complex members may be thought to be derived from groups of interconnected gates. Many of the important characteristics of a family of digital circuits may thus be inferred from the inverting-gate members of that family. The circuit configurations of most NAND and NOR gates follow one of several common designs or "configuration types." The predominant configurations are DCTL direct-coupled-transistor logic (DCTL), resistor-transistor logic (RTL), resistor-capacitor-transistor logic (RCTL), diode-transistor logic (DTL), transistor-transistor logic (TTL), and emitter-coupled-transistor logic (ECTL). These configurations are discussed in a later section of this chapter.

Common Terms. Before proceeding further, we shall list and briefly define a number of commonly used terms that pertain to electrical design functions.

$V_{max\ zero}$. The maximum voltage that will be accepted as a logical 0. In many integrated-circuit designs, $V_{max\ zero}$ is typically 0.6 volt or less.

$V_{min\ one}$. The minimum voltage that will be accepted as a logical 1. $V_{min\ one}$ is typically 1.5 volts or greater.

$I_{in\ zero}$. The maximum current that the input to a circuit will require when the input voltage is at $V_{max\ zero}$.

$I_{in\ one}$. The maximum current that the input to a circuit will require when the input voltage is at $V_{min\ one}$.

$I_{out\ zero}$. The minimum current that the output of a circuit is capable of delivering when the output voltage is at $V_{max\ zero}$.

$I_{out\ one}$. The minimum current that the output of a circuit is capable of delivering when the output voltage is at $V_{min\ one}$.

Fan Out. The number of gate inputs being driven by the output of a similar gate.

Fan-out Capability. The maximum number of gate inputs which may be driven by the output of a similar gate. Fan-out capability N is determined by the relation:

$$\frac{I_{out\ zero}}{I_{in\ zero}} \geq N \leq \frac{I_{out\ one}}{I_{in\ one}} \tag{8-1}$$

Fan In. The number of input terminals of a gate.

Passive-drive State. The logical state at the input of a gate that requires no current flow to maintain that gate. The passive-drive state is indistinguishable from that generated by an open circuit at the input. The opposite state is called the active-drive state.

Characteristic Curves. Since complex functions can be considered as groups of interconnected NAND or NOR gates, it has been found to be convenient to concentrate attention on the simple gate functions when discussing electrical characteristics of a given logic family. Much of the important information related to the input-output characteristics of digital gates can be displayed in two figures.

The voltage-transfer characteristic of an inverting gate is shown in Fig. 8-1. The figure plots the relationship between the voltage on the output and the voltage on one of the inputs, the other inputs being in the passive-drive state.

Two important points are indicated in the figure. These points, the intersections of the limiting input and output voltages, serve as necessary (but not sufficient) acceptance criteria for gates to be used compatibly. For proper operation, the characteristic curve must be above the upper left-hand point and below the lower right-hand point. A gate confined within these limits will have an output voltage in excess of $V_{\text{min one}}$ when $V_{\text{max zero}}$ is applied at the input and an output voltage less than $V_{\text{max zero}}$ when $V_{\text{min one}}$ is applied at the input.

The margins by which the gate characteristic curve passes above the upper left-hand point and below the lower right-hand point are called the 1-state and 0-state noise margins (or noise immunities), respectively. These quantities characterize the relative immunity of the gates to noise pickup. In applications intended for "noisy" environments, such as those found near high-current electromechanical switches, the noise margins are important parameters in the selection of switching circuitry.

Next, consider the current-drive characteristic shown in Fig. 8-2. The

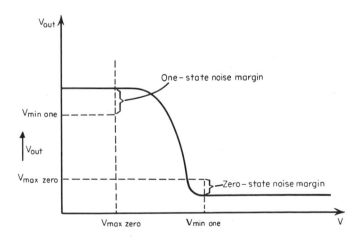

Fig. 8-1   Voltage-transfer characteristic of inverting gate.

figure is characteristic of those circuits having a 0-level active-drive state. Two curves are drawn. One curve is the I_{in} versus V_{in} characteristic (I_{in} is multiplied by N, the fan-out capability). The other is the I_{out} versus the V_{out} characteristic. The intersection of the two curves gives the 0-level voltage at which the input-current capability is sufficient to drive N inputs. For successful operation, the intersection must occur to the left of the $V_{max\ zero}$ line.

Figure 8-3 is the equivalent characteristic for circuits with a 1-level active-drive state. Here the intersection point must occur to the right of the $V_{min\ one}$ line.

Bipolar Integrated-circuit Logic Configurations

Important characteristics in the selection of an integrated-circuit family are power dissipation, speed (as expressed by propagation delay for gates), fan-out capability, and noise margin. These parameters are not independent, and tradeoffs exist among the parameters either in design or use. For example, low values of propagation delay are normally designed at the expense of high power dissipation and vice versa. In fact, the product of power dissipation and propagation delay has been used as a figure of merit in evaluating competitive products. In use, higher noise margins may be obtained by loading circuits with fewer inputs than the specified fan out.

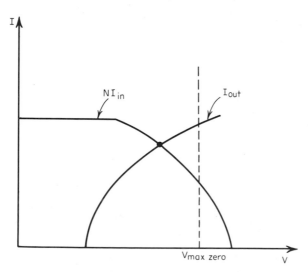

Fig. 8-2   Current-drive characteristic for circuit with 0-level active-drive state.

The step is particularly effective with circuits of the RTL and RCTL types. In qualitatively discussing the pertinent characteristics of the major configuration types below, we have assumed essentially average characteristics and performance.

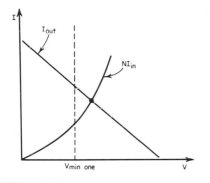

Fig. 8-3 Current-drive characteristic for circuit with 1-level active-drive state.

DCTL (Direct-Coupled-Transistor Logic). The basic circuit configuration for a DCTL gate is shown in Fig. 8-4. Each input is connected directly to the base of a transistor in the common-emitter configuration. The transistors appear in parallel and share a common load resistor R_L

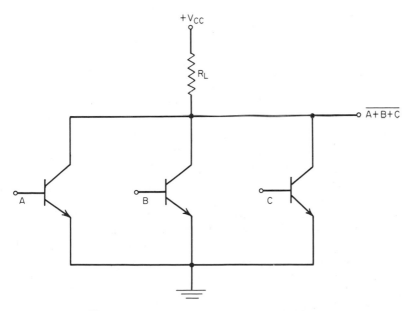

Fig. 8-4 DCTL gate.

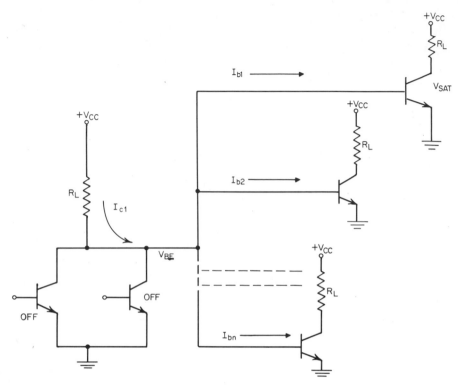

Fig. 8-5   DCTL gate-loading diagram.

which is connected to a positive supply voltage. The passive-drive state for the circuit is the 0 level. The circuit performs the NOR function.*

Logic Voltage Levels. A high voltage at any one of the input terminals will turn on the corresponding transistor and result in a low voltage output. This voltage will be determined by the saturation voltage of the transistor (which must usually be low, for example, 0.2 volt).

If the voltage level at every input is low, none of the transistors will conduct and the output voltage will be high. The circuit will drive one or more inputs similar to its own. The high-level output voltage will be determined by the base-emitter drop of the transistor in the succeeding stage — typically 0.5 volt. Thus, the output voltage varies only about 0.3 volt in going from 0 level to 1 level.

Fan-out Capability. The ultimate fan-out capability of a DCTL gate is determined primarily by the gain of the transistors in the gate. Consider

*This statement assumes a positive logic assignment of voltage levels, i.e., the most positive voltage corresponds to a logical 1. For a negative logic assignment (high voltage = 0 level), the circuit performs the NAND function.

the loading diagram of Fig. 8-5. A gate is shown driving n other gate inputs. The transistors in the driver circuit are off, permitting current to flow into the load circuits. The base current required by each load transistor is

$$I_{bn} = \frac{V_{CC} - V_{SAT}}{R_L} \frac{1}{\beta} \tag{8-2}$$

The collector current available is approximately

$$I_{cl} = \frac{V_{CC} - V_{BE}}{R_L} \tag{8-3}$$

The fan-out capability is therefore

$$N = \frac{I_{cl}}{I_{bn}} = \frac{V_{CC} - V_{BE}}{V_{CC} - V_{SAT}} \beta \approx \beta \tag{8-4}$$

In practice, however, the fan-out capability is considerably less than that indicated by Eq. (8-4). The practical restriction is caused by a phenomenon called "current hogging," arising from normal production tolerances in the transistor base-input $I_b - V_b$ characteristics. These tolerances are depicted by the two curves drawn in Fig. 8-6, representing the $I_b - V_b$ characteristics of two transistors, T_1 and T_2. At the equilibrium driving level $V_{BE(on)}$, transistor T_1 will "hog" the available current and prevent transistor T_2 from obtaining its share. Because the current is not distributed equally between the load transistors, the fan-out capability must be limited.

Evaluation. The DCTL configuration is the simplest and least expensive circuit to fabricate using integrated-circuit technology. Its speed

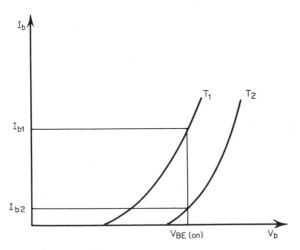

Fig. 8-6 Illustration of current hogging.

capabilities are excellent, being limited only by the transistor storage time, which can be minimized by degrading the lifetime with a gold impurity doping. However, the low-logic voltage swing makes this type particularly vulnerable to errors introduced by electrical noise. Due to its poor noise immunity and restrictions in fan out, the DCTL circuit has not become a popular circuit type.

RTL (Resistor-Transistor Logic). The RTL circuit configuration, shown in Fig. 8-7, is a slightly modified DCTL structure. The addition of resistors at the base inputs equalizes the current distribution to the transistors and thereby improves the fan-out capability. The high-level voltage is increased by the voltage drop across the input resistors. This fact is illustrated by the loading diagram of Fig. 8-8. The high-level voltage (assuming $R_{B1} = R_{B2} = \cdots = R_B$) is found to be

$$
\begin{aligned}
V_{\text{high}} &= \frac{V_{CC} - V_{BE}}{R_L + R_B/n} \frac{R_B}{n} + V_{BE} \\
&= \frac{V_{CC} - V_{BE}}{nR_L/R_B + 1} + V_{BE}
\end{aligned}
\tag{8-5}
$$

where n is the fan out.

The high-level voltage is therefore determined by the fan out. By limiting the fan out, it can be made quite high.

Evaluation. The RTL gate has been a popular logic form, being relatively simple and inexpensive to construct. It has only recently succumbed

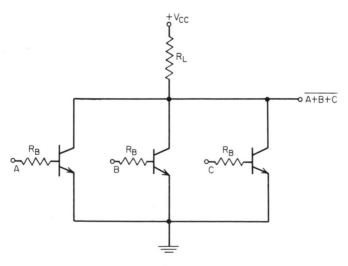

Fig. 8-7   RTL gate.

to the more attractive TTL and ECTL types. The noise immunity can be made relatively high, but only by limiting fan out. The speed of operation is limited by the time constant associated with R_B and its parasitic junction capacitance. By reducing R_B and R_L in value proportionally, better speed characteristics can be obtained, at the expense of increased power dissipation.

RCTL (Resistor-Capacitor-Transistor Logic). The RCTL gate is formed by connecting small "speedup" capacitors across the input resistors of an RTL gate. The speedup capacitors are selected in value to compensate for the stored charge needed by the transistor for conduction. During switching, this charge may be supplied or removed almost instantaneously, thereby enhancing the speed of operation. An RCTL gate is drawn in Fig. 8-9.

Evaluation. The RCTL gate theoretically offers improved speed capabilities over the comparable RTL gate, with considerable sacrifice in simplicity of construction. In practice, this configuration type has seen little use. In one application it has been employed to improve the response of low-power gates. In this low-speed application, however, the resulting operating frequency is still below any of the presently available higher-power RTL gates.

DTL (Diode-Transistor Logic). The DTL gate is essentially a series

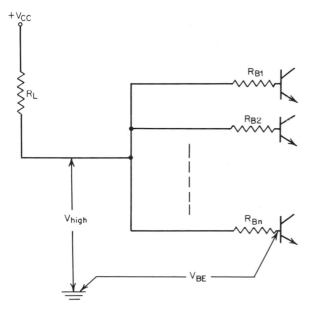

Fig. 8-8   *Loading diagram for RTL gates.*

combination of a diode AND gate and an inverter. The basic circuit configuration is shown in Fig. 8-10. Modifications of this basic configuration are prevalent.

The DTL circuit performs the NAND function. The passive-drive state at the input is a logical 1, i.e., an open circuit at an input acts as a logical 1. The action of the circuit may be understood as follows: Assume first that the voltage applied to all the inputs is at the high level (say at $+ V_{CC}$). The input diodes D_I are back-biased, and current I_2 flows through diodes D_1 and D_2 and into the base of the output transistor (resistor R_D is sufficiently high in value so that it diverts only a small part of I_2). The output transistor is turned on and the output voltage is low. The potential at V_I is the sum of the potential drops across the two offset diodes D_1 and D_2 and the $V_{BE(on)}$ of the output transistor. The offset diodes thus act in a voltage-level-setting capacity. The node voltage V_I is typically 1.5 to 2.0 volts under these conditions. Now imagine that the voltage at one of the input terminals is gradually lowered. At some point, one of the diodes D_I becomes forward-biased and begins to conduct current I_1. As the input voltage is reduced still further, the available current favors the I_1 branch, and the I_2 goes to 0. The output transistor turns off, and the output voltage goes to the high level.

Logic Voltage Levels. The low-level voltage is determined by the saturation voltage of the output transistor. This voltage is typically 0.2 to 0.5 volt, depending on the fan out (current from the load circuits goes through the output transistor to ground, raising the saturation voltage).

At the output logical-1 state, no current flows through resistor R_u, and

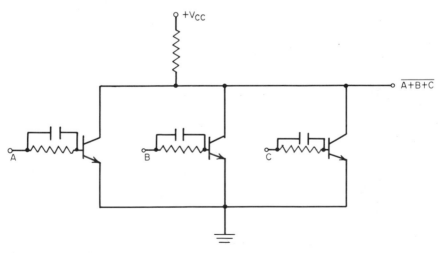

Fig. 8-9   RCTL gate.

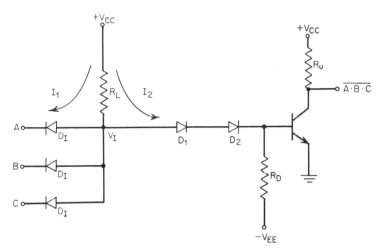

Fig. 8-10   DTL gate.

the voltage level therefore equals V_{CC} (typically $+4$ volts). The DTL circuit therefore has well-separated voltage levels and a resulting good noise immunity.

The resistor R_u (called the pull-up resistor) may be deleted from the circuit since load is provided by the R_L resistors of the load circuits. This step reduces the power consumption and increases the fan-out capability by one, but it also reduces the logic-level swing. Refer to the loading diagram of Fig. 8-11. Assume transistor T_1 is biased off. Without R_u connected, diode D_I cannot be back-biased. The logical-1 output voltage V_O is determined by voltage V_I minus the forward drop across diode D_I (D_I conducts

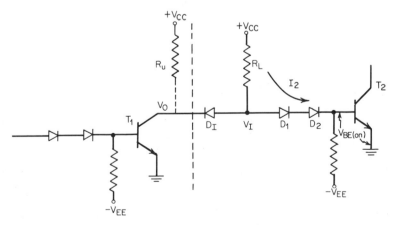

Fig. 8-11   Loading diagram for DTL gate.

the small leakage current required by the collector of transistor T_1). V_O is thus typically in the neighborhood of 1.0 to 1.5 volts.

Resistor R_D provides a current-drain path for stored charge in the transistor during the turnoff period. The resistor may be returned to ground or a negative voltage V_{EE}, the latter connection providing improved operating speed.

Fan-out Capability. The fan-out capability of a DTL gate is determined by the saturation-voltage characteristic of the output transistor. Loads may be added until this transistor can no longer support the current required while maintaining a low output voltage. The fan-out capability is generally higher than that of the RTL circuit type.

Evaluation. The DTL circuit, often called the workhorse of integrated-circuit designs, is a widely used circuit type. It has a good speed characteristic, excellent noise immunity (nearly independent of fan out), and a good fan-out capability. On the debit side is the circuit complexity and somewhat higher power consumption.

TTL (Transistor-Transistor Logic). The TTL (or T²L) circuit was developed from the basic DTL configuration. A diagram of this circuit type is shown in Fig. 8-12.

The TTL gate replaces the gating diodes and one of the two offset diodes of the DTL gate with a multiemitter transistor. This input transistor is always in the saturation region of operation with a relatively constant base current I_B. I_B is steered either to the base of the output transistor or out through one or more of the input emitters. Since the input-transistor base current remains relatively constant, little charge adjustment is required during switching, and switching speeds are considerably improved. Furthermore, the input transistor provides a fast sink for stored charge in the output transistor during the turnoff transient.

Some price is paid for the improved switching speeds. First, the lack of

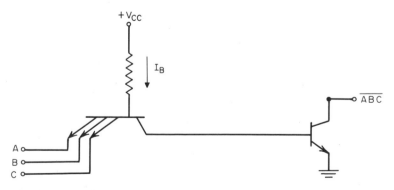

Fig. 8-12   TTL gate.

an offset diode results in a lower noise margin than that of a comparable DTL gate. Second, a current-hogging problem similar to that of the DCTL configuration exists, limiting the fan out. For example, consider the load situation of Fig. 8-13. The output transistor of a driving circuit T_1 is connected to the inputs of two or more similar gates. Assume T_1 is off and therefore T_4 and T_5 are conducting. Suppose that V_{BE_5} is less than V_{BE_4}, either through nonequal collector currents I_{C_4} and I_{C_5} or because of production tolerances. Transistor T_3, operating in the inverse direction, will "pull" current I_1 from the emitter of T_2. The base drive of T_4 is reduced by this amount. If more than two inputs are tied together, the problem is correspondingly compounded. Current hogging can be minimized by designing the input transistors to have a very low inverse gain.

Evaluation. The TTL configuration is a simple circuit with excellent speed characteristics and an acceptable but somewhat reduced noise immunity and fan-out capability as compared to the DTL configuration. It is currently very much in favor and widely used.

ECTL (Emitter-Coupled-Transistor Logic). All the previously discussed integrated-circuit types employ transistors that are allowed to

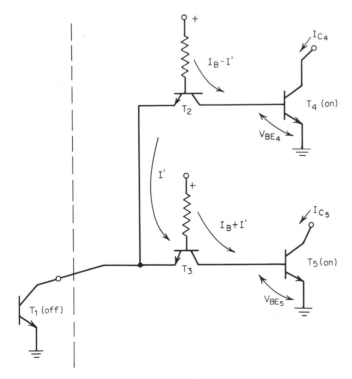

Fig. 8-13   Current hogging in TTL gates.

saturate when turned on. A saturated transistor stores an excess charge which must be removed during the turnoff transient. This requirement limits the speed of operation.

The ECTL configuration employs transistors that operate only in the active region. These circuits therefore have the fastest speed characteristic of those integrated circuits presently available.

The basic configuration of an ECTL gate is shown in Fig. 8-14. The input transistors are emitter-coupled to a reference transistor. The resistor at the emitter terminal provides a negative feedback and a resultant high input-impedance characteristic. All transistors are quiescent-operating. A high-level voltage at any input will result in an increased current I_1 being drawn which tends to raise the potential V_E. The latter change tends to turn the reference transistor off, reducing the current I_2. The sum of I_1 and I_2 remains essentially constant.

A typical ECTL circuit has two outputs with both the NOR and OR functions available, as shown in the figure. The difference between the logical-0 and logical-1 levels is normally small (0.4 volt), but the noise immunity is moderately good due to the feedback in the circuit. Fan-out capability is very high because of the high impedance of the input transistors.

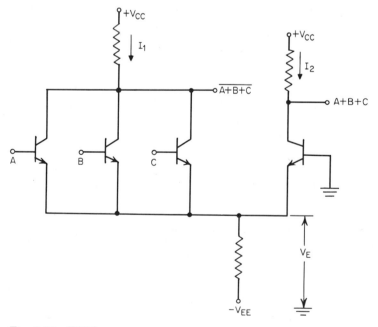

Fig. 8-14 ECTL gate.

Evaluation.   The ECTL configuration is a superior circuit in speed and fan out. Noise immunity is comparable to that of the TTL and RTL circuits. Power consumption, as well as complexity (and hence cost), is high. In addition, the negative reference voltage must be highly stable and must usually be temperature-compensated.

Wired Logic.   Under certain circumstances, the output terminals of two or more inverting gates may be connected to expand fan-in capability or to achieve an additional level of logic. Connecting output terminals to a common node has important ramifications for the electrical operation of the circuits, and careful consideration must be given to the possibility of introducing deleterious effects. Where such interconnection is permitted, manufacturers generally publish guidelines which should be strictly followed.

In DCTL, RTL, and RCTL circuit configurations, interconnecting the outputs of two or more gates results in a single NOR gate with increased fan-in capability. Fig. 8-15 illustrates the interconnection of two RTL gates. Note that the gating transistors are placed in parallel, thus effectively doubling the fan-in capability. Unfortunately, the load resistors are also placed in parallel, requiring any single transistor to conduct twice the normal current when in the "on" state. To avoid this effect, the positive power-supply voltage can be connected to only one of the interconnected gates, or special gates may be constructed without load resistors.

In DTL and TTL circuits, interconnection of the outputs results in an additional level of logic. The output from the common node is the AND

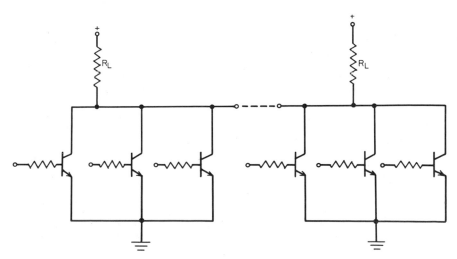

Fig. 8-15   Interconnection of two RTL gates to expand fan-in capability.

function of the individual outputs. Since the additional logic capability is attained through wiring without increasing the number of circuits, it is referred to as "wired logic" or "collector logic."

Figure 8-16a illustrates the interconnection of two DTL gates. The equivalent logic function is drawn in Fig. 8-16b.

Use of Wired Logic. The wired AND function adds processing power to NAND circuitry and may, in many cases, reduce the number of gates required to achieve a given logical function. As an example of its use, consider the circuitry required to achieve the "exclusive OR" function. The normal implementation using NAND gates requires three gates while the

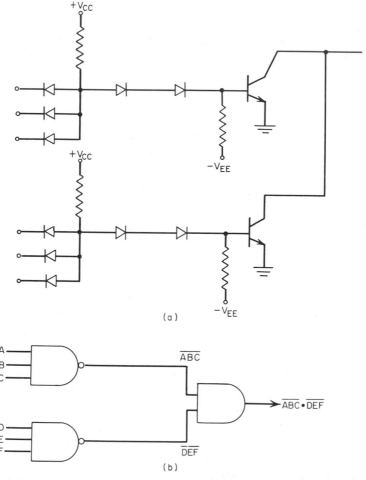

Fig. 8-16 Interconnection of two DTL gates. (a) Schematic and (b) equivalent function to achieve wired AND logic.

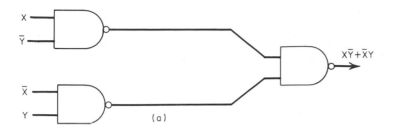

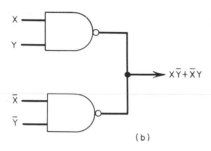

Fig. 8-17 *"Exclusive OR" function. (a) Conventional NAND implementation. (b) NAND implementation using wired AND function.*

version employing a wired AND function requires only two gates. The two configurations are compared in Fig. 8-17. The wired AND function is represented with a small circle in the figure.

The basic design model of the wired AND configuration is identical to that of the two-level NAND design with the exception of the final inversion of the output of the latter. This similarity provides the key for systematically designing functions using a wired AND configuration, namely, that we must first complement the desired output function and then proceed with the normal two-level NAND design rules. For example, consider the expression

$$f = ABC + A\bar{B}\bar{C} + \bar{A}BC + A\bar{B}C$$

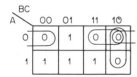

This function is plotted in the Karnaugh map of Fig. 8-18. We simplify the complement of f (a procedure equivalent to circling the 0s), to obtain

Fig. 8-18 *Plot of sample function.*

$$\bar{f} = \bar{A}B + \bar{A}\bar{C} + B\bar{C}$$

The function f is then obtained with three gates using a wired AND output. The inputs to the first gate are $\bar{A}$ and B, to the second $\bar{A}$ and C, and to the third B and $\bar{C}$. The wired AND version is compared with the normal NAND implementation in Fig. 8-19.

MOSFET Digital Integrated Circuits

MOSFET integrated circuits currently have a small but growing share in the digital integrated-circuit market. There are a number of digital-system applications, particularly those requiring low power consumption, in which MOSFET integrated circuits are presently used to great advantage.

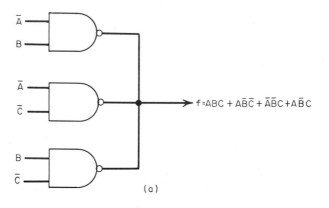

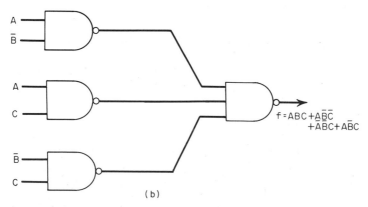

Fig. 8-19   *Two implementations of sample function.* (a) *Wired AND implementation.* (b) *Normal NAND implementation.*

The outlook for these circuits is quite promising. The basic constructional simplicity of the MOSFET and the fact that it permits large-scale integration — where large and often complex arrays are fabricated on a single small chip — provide a processing power-to-cost ratio that may be unsurpassed by any other digital-circuit approach. It may be anticipated that the potential of MOSFET digital integrated circuits will be fully exploited in the years to come.

The MOSFET in Digital Applications. The coupling compatibility of circuit stages is of first importance in the selection of a MOSFET type for digital applications. The simplest and most desirable method of coupling two stages is to couple them directly, i.e., with the output of one MOSFET connected directly to the input of another. Thus the MOSFETs used must be dc-compatible with respect to input and output voltage ranges. The normally nonconducting enhancement-mode devices meet this requirement. (The n-channel enhancement-mode MOSFET operates with a positive drain-to-source potential and a positive gate-to-source potential. The p-channel enhancement-mode MOSFET operates with negative drain and gate voltages.) On the other hand, depletion-mode devices require gate and drain voltages of opposite polarity. Thus MOSFETs employed in digital circuits are almost exclusively enhancement-mode types.

Either n- or p-channel enhancement-mode MOSFETs may be employed. Often both appear in the same circuit. The n-channel MOSFET has a somewhat higher theoretical operating speed than the p-channel MOSFET because the mobility of electrons is higher than that of holes. However, in the current state of the technology, p-channel enhancement-mode MOSFETs can be constructed with somewhat greater success and are thus favored. It will be recalled that the surface-state charge tends to invert the surface toward n-type. In n-channel enhancement-mode devices, the p-type substrate must be more heavily doped to compensate for the surface inversion in order to prevent the formation of a preexisting n channel (it would otherwise become a depletion-mode device). The drain-to-substrate breakdown voltage is subsequently lower in such devices.

Although relatively fast (UHF) MOSFETs may be constructed, the high-impedance characteristic of MOSFETs makes them quite susceptible to the speed-degrading effects of capacitance. The operating speed of MOSFET digital integrated circuits is thus generally determined by the capacitance at the various circuit nodes that must be charged or discharged during switching. The capacitance associated with the physical package, i.e., the header and lead capacitance, can present a particular obstacle to the attainment of high switching speeds. There is thus another reason for tending toward large-scale integration of MOSFET arrays. The internal nodes of such arrays are not affected by header capacitance, and the complex internal function can be performed considerably faster than in the individual wired-circuit counterpart. The fastest digital-MOSFET arrays can operate about one-tenth as fast as the fastest equivalent bipolar arrays.

The high impedance of MOSFETs can be used to advantage in constructing low-power circuits. A number of digital MOSFET circuits have been designed to consume power only during the transient switching period, requiring a negligible small current drain to maintain a logical state.

Typical Configurations. MOSFET digital integrated circuits are composed of groups of interconnected inverters and transmission gates. The appropriately wired inverters perform logical operations and the transmission gates are used for transferring information at controlled intervals. Timing is often attained through use of two or more clock-pulse trains with phased (sequentially occurring) clock-pulse intervals.

MOSFET Inverter. An inverter circuit using two p-channel MOSFETs is shown in Fig. 8-20. Transistor T_2 acts as a load resistor for switching transistor T_1. When the magnitude of the input voltage V_{in} is below the threshold voltage magnitude, transistor T_1 is nonconducting and the output voltage V_{out} approaches the negative supply voltage V_{DD}. When the input voltage exceeds V_{th} in magnitude, transistor T_2 is conducting and the magnitude of V_{out} drops. With a high magnitude of V_{in} it is desirable for V_{out} to be close to ground potential. Qualitatively speaking, the potential drops across the two transistors under this condition are proportional to

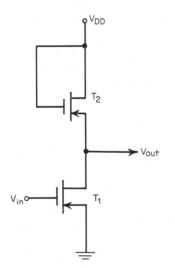

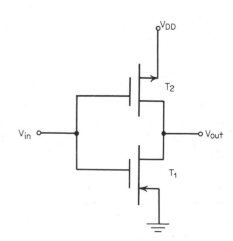

Fig. 8-20   Inverter circuit using two p-channel MOSFETs.

Fig. 8-21   Complementary inverter circuit.

their respective channel resistances. To obtain a low value of V_{out}, the channel width of T_1 is normally constructed to be considerably greater than that of T_2, the transistors being otherwise identical. The effective channel resistance of T_1 is thus made to be small with respect to that of T_2, and a low magnitude of V_{out} is thereby assured. It can be shown (see Prob. 8-1) that with an applied input voltage equal to V_{DD}, the resulting output voltage is

$$V_{out} = (V_{DD} - V_{th})\left[1 - \left(\frac{w_1}{w_1 + w_2}\right)^{1/2}\right] \tag{8-6}$$

where w_1 and w_2 are the channel widths of transistors T_1 and T_2 respectively.

An alternative inverter circuit using both p- and n-channel MOSFETs is shown in Fig. 8-21. With V_{in} close to ground potential, transistor T_1 is off and T_2 is on. The output voltage is therefore close to V_{DD}. When V_{in} is close to V_{DD}, T_1 is on and T_2 is off, resulting in an output potential close to ground. It should be noted that the complementary inverter circuit has no steady-state current flow. The complementary feature also improves the operating speed and allows both transistors to be made as physically small as technology permits. A disadvantage is the constructional complexity of fabricating both n- and p-channel devices on the same chip.

MOSFET Transmission Gate. The MOSFET transmission gate is employed to achieve controlled coupling between two circuit nodes. The transmission gate is illustrated in Fig. 8-22. When the magnitude of the gate potential V_G is less than the threshold-voltage magnitude, the output is isolated from the input. Upon raising the gate-potential magnitude such that $V_G > V_{in} + V_{th}$, the output is coupled to the input and $V_{out} = V_{in}$.

MOSFET Logic Gates. An endless variety of logic configurations can be constructed using variations of the MOSFET inverters just discussed. Common to all is an exceptionally high fan-out capability. Since the gate-input terminals do not conduct current, there are no dc limitations on fan out. The fan out is limited only by the accumulative effects of input

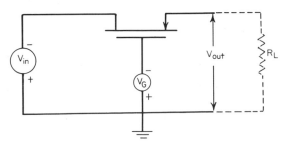

Fig. 8-22 MOSFET transmission gate.

capacitance on operating speed. The high input impedance also eliminates current-hogging problems.

Two typical configurations are presented here. In Fig. 8-23 is shown a negative-logic NOR gate similar in configuration to the DCTL structure previously discussed. A negative-logic NAND gate employing complementary transistor pairs is shown in Fig. 8-24. The switching transistors are series-connected. The output can go toward ground only when all switching transistors are in the conducting state.

MOSFET Flip-flops. MOSFET flip-flops are constructed by cross-coupling inverting gates. Clocked flip-flop types such as the *RS* and *JK* varieties are normally implemented using the asynchronous design techniques discussed in the previous chapter.

The high-impedance characteristic of the gate input permits long-term storage of voltage levels on the input-gate capacitance. A profitable use of this phenomenon is illustrated by the shift register of Fig. 8-25. The shift register employs inverters (*I*) separated by transmission gates (*T*). Alternate transmission gates are strobed with one of two clock-pulse streams, phased apart. A set of two inverters and two transmission gates forms a stage of the shift register. At clock-pulse time ϕ_1, the state of the first inverter in each stage is transmitted to the input of the second inverter of that stage. The signal can go no further, being blocked by the transmission gates strobed by ϕ_2. When ϕ_1 goes off, the levels remain stored on the input capacities of the second inverters. At clock-pulse time ϕ_2, the signals are

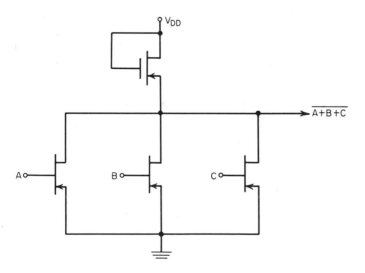

Fig. 8-23   MOSFET NOR gate.

transmitted to the first inverters of the succeeding stages, and the process is repeated. This simple and economical implementation of shift registers has found wide application in the technology.

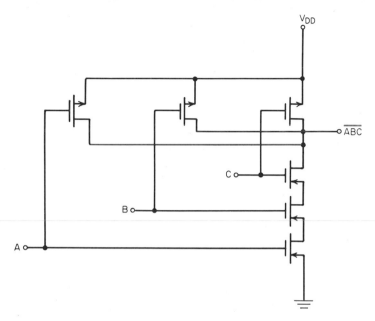

Fig. 8-24   MOSFET NAND gate.

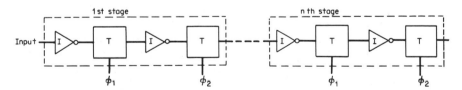

Fig. 8-25   MOSFET shift register.

PROBLEMS

8-1. Assume that the transistors of Fig. 8-20 are identical except for channel widths, and that $V_{in} = V_{DD}$. Under these conditions T_2 is saturated and T_1 is unsaturated. Verify Eq. (8-6).

NINE

Linear Integrated Circuits

The theory fundamental to linear-circuit design is extensive and cannot be suitably treated either in scope or depth in this text. We adopt instead some coverage of topics of special importance to linear integrated circuits.

In this chapter, the transistor is considered as a linear amplifying device. A particularly useful amplifier configuration for linear integrated circuits is the differential amplifier, and the properties of this popular and basic configuration are reviewed. The fabrication of highly functional linear integrated circuits was not an easy task for integrated-circuit workers and required ingenious application of new design methods. Some of these techniques are considered before passing to the final chapter topic, dealing with externally applied feedback and compensation.

Background

Linear amplifiers are designed to operate in the relatively narrow region between the extreme full-on and full-off operating points that are characteristic of digital circuits. The input-voltage range between these two points depends upon the gain of the amplifier. A high-gain amplifier will have a narrow operating region because small changes in input voltage will result in large changes in output voltage.

It is necessary to initially bias the amplifier to a quiescent point close to the middle of the operating range and to maintain that quiescent point throughout the operating life of the amplifier. The latter requirement can be difficult to achieve. The values of component parts, such as resistors, change with temperature and with age, and such changes may be equated to the insertion of small dc voltages within the circuit. If the equivalent voltages are close to the input, they may be amplified by the circuit to drive it beyond the linear operating region. Thus linear circuits are quite sensitive to component-part variations and require tighter tolerances and lower temperature coefficients than digital circuits.

In conventional circuit design practice, linear circuits are often customized to the specific application. In ac applications, for example, individual amplifier stages are normally capacitively coupled to prevent the propagation and further amplification of dc disturbances arising in a given amplifier stage. Often, large capacitors in the tens of microfarads are required to maintain frequency response at the low-frequency end, such as in audio and pulse amplifiers. Alternatively, transformer coupling between stages can be used. Critical circuit sections of dc amplifiers may be constructed with precision and low-temperature-coefficient component parts.

The aim in dealing with integrated circuits is to design general-purpose amplifiers for a multiplicity of applications. The necessary coupling, feedback, and compensation as dictated by the application are employed at the amplifier level rather than at the amplifier-stage level. Little would be gained in cost reduction, miniaturization, or reliability by supplying single-stage integrated circuits with few component parts.

Difficulties arise when one attempts to implement the design of an entire integrated-circuit amplifier. Conventional design steps are not open to the integrated-circuit fabricator. Practical integrated-circuit component parts have poor tolerances, high temperature coefficients, and limited ranges of realizable values. The use of precision parts, inductors, or high-valued coupling capacitors is out of the question.

These deficiencies must be accepted, and new techniques using the advantages of integrated-circuit technology employed. There are a number of special attributes of integrated-circuit component parts that may be used to advantage. Among these are:

1. The relatively low cost of high-quality transistors

2. The exceedingly good matching of the characteristics of physically adjacent transistors

3. The tight tolerance of resistance ratios which can be maintained in spite of temperature variations

The clever use of these and other attributes of the technology has made possible the development of the present versions of linear integrated circuits. These are entirely competitive in characteristics and cost with discrete-component circuits.

Linear-circuit Properties of Bipolar Transistors

In linear-circuit design, the transistor is often described as a linear two-port network. The parameters of this network are called the small-signal parameters. In general, a transistor is far from being a linear device. However, at a constant quiescent operating point with small-signal excursions, the transistor may be approximated by a linear model.

Additional considerations must be given to the behavior of the transistor as the signal frequencies are increased. The model must now include reactive elements which lead to decreased amplification and signal phase shifts.

Two-port Network

Low-frequency Characteristics. A transistor two-port network is shown in Fig. 9-1. The common-base configuration is used for this example. The network is described by four variables—the emitter-base and collector-base voltages and the emitter and collector currents. The lower-case symbols denote incremental or ac quantities. As a first approximation, we let the network contain three elements—the input resistance r_{ib}, the current generator $g_m v_{eb}$ (g_m is the forward transconductance), and the output resistance r_{ob}. Note that we could just as well have represented the current generator as αi_e (corresponding to selecting i_e rather than v_{eb} as the independent variable).

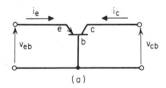

(a)

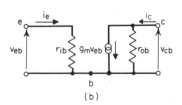

(b)

Fig. 9-1   (a) Transistor common-base configuration. (b) Equivalent two-port network.

To determine the parameters of the model, we refer to the Linvill equations for transistor operation [see Prob. 2-5 or Eqs. (3-30), modified for assumed direction of current flow]:

$$I_E = \left(H_D + H_{C_1} + S_1 \frac{d}{dt} \right) \hat{p}_E - H_D \hat{p}_C$$

$$I_C = -H_D \hat{p}_E + \left(H_D + H_{C_2} + S_2 \frac{d}{dt} \right) \hat{p}_C \tag{9-1}$$

The current terms in these equations represent the total current values, i.e., steady-state plus incremental values. The excess minority-carrier densities may be expressed in terms of the emitter-base and collector-base voltages as

$$\hat{p}_E = p_{n_0} \left(\exp \frac{eV_{EB}}{kT} - 1 \right)$$

$$\hat{p}_C = p_{n_0} \left(\exp \frac{eV_{CB}}{kT} - 1 \right) \tag{9-2}$$

Combining Eqs. (9-1) and (9-2), and setting $d/dt = s$ (the Laplace operator), we obtain an alternate formulation describing transistor operation:

$$I_E = (p_{n_0} H_D + p_{n_0} H_{C_1} + p_{n_0} S_1 s) \left(\exp \frac{eV_{EB}}{kT} - 1 \right)$$

$$- p_{n_0} H_D \left(\exp \frac{eV_{CB}}{kT} - 1 \right)$$

$$I_C = -p_{n_0} H_D \left(\exp \frac{eV_{EB}}{kT} - 1 \right) + (p_{n_0} H_D + p_{n_0} H_{C_2}$$

$$+ p_{n_0} S_2 s) \left(\exp \frac{eV_{CB}}{kT} - 1 \right) \tag{9-3}$$

Equations of this basic form were first devised by Ebers and Moll,[*] and may be expressed in terms of measurable quantities as

$$I_E = -\frac{I_{EO}}{1 - \alpha_F \alpha_I} \left(1 + \frac{s}{\omega_F} \right) \left(\exp \frac{eV_{EB}}{kT} - 1 \right)$$

$$+ \frac{\alpha_I I_{CO}}{1 - \alpha_F \alpha_I} \left(\exp \frac{eV_{CB}}{kT} - 1 \right)$$

$$I_C = \frac{\alpha_F I_{EO}}{1 - \alpha_F \alpha_I} \left(\exp \frac{eV_{EB}}{kT} - 1 \right) \tag{9-4}$$

$$- \frac{I_{CO}}{1 - \alpha_F \alpha_I} \left(1 + \frac{s}{\omega_I} \right) \left(\exp \frac{eV_{CB}}{kT} - 1 \right)$$

[*] J. J. Ebers and J. L. Moll, "Large-Signal Behavior of Junction Transistors," *Proc. IRE*, vol. 42, no. 12, pp. 1761–1772, December, 1954.

where α_F = forward dc common-base current gain

$\quad\quad \alpha_I$ = inverse dc common-base current gain (i.e., transistor operated with collector acting as emitter and vice versa)

$\quad I_{CO}$ = collector leakage current with emitter open

$\quad I_{EO}$ = emitter leakage current with collector open

$\quad\quad \omega_F$ = forward radian cutoff frequency (common base)

$\quad\quad \omega_I$ = inverse radian cutoff frequency (common base)

It may be shown that the following relationships exist between the parameters of the two models represented by Eqs. (9-3) and (9-4):*

$$p_{n_0}H_D = \frac{-\alpha_F I_{EO}}{1 - \alpha_F \alpha_I}$$

$$p_{n_0}H_{C_1} = \frac{-I_{EO}(1 - \alpha_F)}{1 - \alpha_F \alpha_I}$$

$$p_{n_0}H_{C_2} = \frac{-I_{CO}(1 - \alpha_I)}{1 - \alpha_F \alpha_I} \tag{9-5}$$

$$p_{n_0}S_1 = \frac{-I_{EO}}{(1 - \alpha_F \alpha_I)\omega_F}$$

$$p_{n_0}S_2 = \frac{-I_{CO}}{(1 - \alpha_F \alpha_I)\omega_I}$$

For the purposes of our example, a number of simplifications may be made on Eqs. (9-4). In the typical amplifier configuration, the collector-base junction is reverse-biased, and V_{CB} is a negative voltage which is large with respect to kT/e. The exponential terms involving V_{CB} are therefore a negligible contribution to the total currents. The emitter-base junction is forward-biased, and the exponential terms involving this quantity are predominant. Also, at low frequencies ($\omega \ll \omega_F$), the time-dependent terms (those involving $s = d/dt$) may be deleted. Making these simplifications, we are left with

$$I_E \approx \frac{-I_{EO}}{1 - \alpha_F \alpha_I} \exp \frac{eV_{EB}}{kT}$$

$$I_C \approx \frac{\alpha_F I_{EO}}{1 - \alpha_F \alpha_I} \exp \frac{eV_{EB}}{kT} \tag{9-6}$$

The small-signal input resistance r_{ib} is defined as the ratio of the incre-

---

*Linvill, *Models of Transistors and Diodes*, p. 140.

mental value of input voltage to the incremental value of input current. Thus, using the first of Eqs. (9-6), we obtain

$$r_{ib} = \frac{v_{eb}}{i_e} = \frac{\partial V_{EB}}{\partial I_E} = \left(\frac{\partial I_E}{\partial V_{EB}}\right)^{-1}$$
$$= \left(\frac{e}{kT} \frac{-I_{EO}}{1 - \alpha_F \alpha_I} \exp \frac{eV_{EB}}{kT}\right)^{-1} = \frac{kT}{eI_E} \tag{9-7}$$

This result shows that the small-signal input resistance is simply the reciprocal of the emitter current times the thermal voltage kT/e. At room temperature, $kT/e = 0.025$ volt, and we have (for I_E expressed in milli-amperes)

$$r_{ib} = \frac{25}{I_E} \tag{9-8}$$

The above is the intrinsic resistance, i.e., the resistance directly attribut-able to transistor action. Later we will want to add extrinsic terms, e.g., the ohmic drop in the base region.

The small-signal forward transconductance g_m is defined as the ratio of the incremental value of collector current to the incremental value of input voltage. From the second of Eqs. (9-6), we obtain

$$g_m = \frac{i_c}{v_{eb}} = \frac{\partial I_C}{\partial V_{EB}} = \frac{e}{kT} \frac{\alpha_F I_{EO}}{1 - \alpha_F \alpha_I} \exp \frac{eV_{EB}}{kT} = \frac{e}{kT} I_C \tag{9-9}$$

The transconductance is therefore proportional to the collector current, the proportionality constant being the reciprocal of the thermal voltage. We may check that our result is consistent with the equality $g_m v_{eb} = \alpha_F i_e$ by noting that

$$g_m v_{eb} = g_m r_{ib} i_e = \frac{e}{kT} I_C \frac{kT}{eI_E} i_e$$
$$= \frac{I_C}{I_E} i_e = \alpha_F i_e$$

The small-signal output resistance r_{ob} is defined as the ratio of the in-cremental value of the collector-base voltage to the incremental value of the collector current. We note that the simplified Eqs. (9-6) do not contain V_{CB} as a variable, and we must therefore use the more complete set of Eqs. (9-4). If recourse to these equations is had, one determines that the output conductance is negligibly small (very large output resistance). Thus the intrinsic output resistance may be neglected. There is, however, a second-order effect which does result in a nonnegligible output conduct-ance.* It arises from the modulating effect of collector-base voltage on

---

*This phenomenon is called the Early effect. See J. M. Early, "Effects of Space-Charge Layer Widening in Junction Transistors," *Proc. IRE*, vol. 40, no. 11, pp. 1401–1406, November, 1952.

the depletion width of the collector-to-base junction, and hence upon the effective base width. The variations in base width, in turn, vary the transistor gain and hence the output current. This relationship may be defined from the following equations (assuming constant I_E):

$$r_{ob} = \frac{\partial V_{CB}}{\partial I_C} = \left(\frac{\partial I_C}{\partial V_{CB}}\right)^{-1} = \left[\frac{\partial(\alpha_F I_E)}{\partial V_{CB}}\right]^{-1}$$

$$= \left(I_E \frac{\partial \alpha}{\partial V_{CB}}\right)^{-1} = \left(I_E \frac{\partial \alpha}{\partial w_B} \frac{\partial w_B}{\partial V_{CB}}\right)^{-1}$$

where w_B = base width.

The output resistance is sufficiently high that it may be neglected in many instances.

We may now easily determine the small-signal parameters for the common-emitter transistor configuration. The small-signal linear two-port equivalent circuit for this configuration is shown in Fig. 9-2. The input resistance r_{ie} is defined as the ratio of value of the base-to-emitter voltage to the incremental value of the base current:

$$r_{ie} = \frac{v_{be}}{i_b} = \frac{\partial V_{BE}}{\partial I_B} = \frac{\partial V_{EB}}{\partial I_E} (1 + \beta_F) = r_{ib} (1 + \beta_F) \tag{9-10}$$

where β_F = low-frequency common-emitter forward current gain.

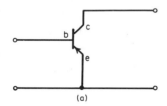

(a)

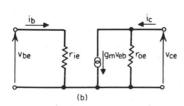

(b)

Fig. 9-2 **(a) Basic configuration and (b) linear-equivalent network for common-emitter configuration.**

In the above equation, we have made use of the relation $I_E = -(I_B + I_C) = -(I_B + \beta_F I_B) = -(1 + \beta_F)I_B$. The common-emitter resistance is larger than the common-base input resistance by a factor approximately equal to the common-emitter current gain β_F.

The transconductance is the same as for the common base [Eq. (9-9)] except for a reversal of sign due to the sign difference between v_{bc} and v_{eb}. The output resistance r_{oe} may be shown to be approximately equal to r_{ob}.

To complete the low-frequency equivalent circuits, we may now add the extrinsic terms consisting of nonidealized ohmic drops at the terminals and within the transistor bulk. Of these, the ohmic drop in the base (and, for integrated-circuit transistors, in the collector) can be of significance. These terms are represented by r_{bb} and r_c in the more complete linear-equivalent circuits of Fig. 9-3. In the figure, the directions of current flow are adjusted so that all terms developed above may be treated as positive quantities. Note that the extrinsic base resistance contributes directly to the input resistance in the common-emitter case, but adds only the term $r_{bb}/(1 + \beta_F)$ to the input resistance of the common-base configuration (since the current flow through r_{bb} is only the fraction $1/(1 + \beta_F)$ of the emitter current).

High-frequency Characteristics. At frequencies that are not small with respect to the cutoff frequency, the effects of the reactive elements of a transistor are significant. We consider here additions appropriate to the linear-equivalent circuits just derived to extend their validity to the high-frequency regime. Returning to the first of Eqs. (9-4), we note that since time dependence is now important, the term involving s/ω_F must be in-

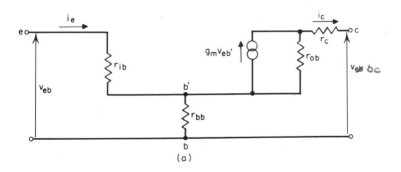

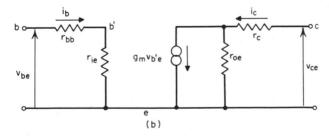

Fig. 9-3 *Low-frequency linear-equivalent networks.* **(a)** *Common base and* **(b)** *common emitter, including extrinsic resistance terms.*

cluded in the analysis. The simplified transistor equation corresponding
to the first of Eqs. (9-6) is

$$I_E \approx -\frac{-I_{EO}}{1 - \alpha_F \alpha_I}\left(1 + \frac{s}{\omega_F}\right)\exp\frac{eV_{EB}}{kT} \tag{9-11}$$

Consider now the input impedance. Taking the differential of both sides
of Eq. (9-11), we obtain

$$dI_E \approx \frac{e}{kT}\frac{-I_{EO}}{1 - \alpha_F \alpha_I}\left(1 + \frac{s}{\omega_F}\right)\exp\left(\frac{eV_{EB}}{kT}\right)dV_{EB} \tag{9-12}$$

For sufficiently small incremental changes in V_{EB}, the exponential term
may be treated as a constant with time, and we may equate the differential
terms with their corresponding incremental values. The revised expression
becomes

$$i_e \approx \left(\frac{e}{kT}\frac{-I_{EO}}{1 - \alpha_F \alpha_I}\exp\frac{eV_{EB}}{kT}\right)\left(1 + \frac{s}{\omega_F}\right)v_{eb} \tag{9-13}$$

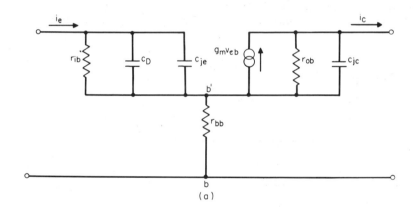

(a)

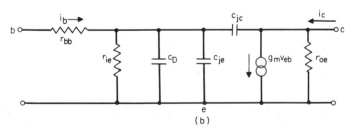

(b)

**Fig. 9-4 High-frequency linear-equivalent networks. (a)
Common base and (b) common emitter.**

The first expression in parentheses is recognized as the reciprocal of the input resistance found previously. For sinusoidal v_{eb}, we may equate the Laplace operator s with the complex frequency term $j\omega$. The complex impedance is therefore found to be

$$z_{ib} = \frac{v_{eb}}{i_e} = \frac{r_{ib}}{1 + \dfrac{j\omega}{\omega_F}} \qquad (9\text{-}14)$$

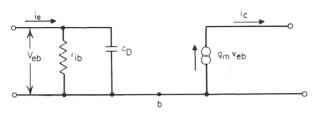

Fig. 9-5 *Simplified high-frequency-equivalent circuit for common-base configuration.*

Equation (9-14) is precisely the impedance expression for a parallel network consisting of a resistor r_{ib} and a capacitor $C_D = 1/r_{ib} \ \omega_F = eI_E/kT\omega_F$. This intrinsic capacitance is usually called the transistor-diffusion capacitance. A revised equivalent circuit for the common-base configuration is shown in Fig. 9-4a. Extrinsic capacitance terms C_{je} and V_{je}, representing the emitter-base and collector-base junction capacities, respectively, have also been included in the figure. Note that they are connected to the intrinsic base node b' rather than to the outside terminal base node b. In the approximate equivalent circuit, the base-spreading resistance r_{bb} represents the resistance between the base contact and the ideal intrinsic base node.

The high-frequency linear-equivalent network for the common-emitter configuration is shown in Fig. 9-4b. The diffusion capacitance C_D has the same value as in the common-base configuration (see Prob. 9-1).

The addition of the reactive elements to the linear-equivalent networks allows us to determine approximately the frequency-dependent behavior of transistors. In particular, we will find that the effective transistor gain decreases as frequency increases and the current experiences a net phase shift in traversing the transistor which increases as frequency increases. The qualitative aspects of these phenomena for the common-base configuration may be adequately illustrated by the simplified equivalent circuit of Fig. 9-5, in which are included only the input resistance r_{ib}, the diffusion capacitance C_D, and the current generator $g_m v_{eb}$. We seek to determine the collector current as a function of frequency with a constant magnitude of emitter current applied. We note

$$v_{eb} = \frac{i_e}{1/r_{ib} + j\omega C_D} \tag{9-15}$$

so that

$$i_c = g_m v_{eb} = \frac{g_m r_{ib}}{1 + j\omega r_{ib} C_D} \, i_e \tag{9-16}$$

But

$$g_m r_{ib} = \frac{e}{kT} \, I_C \frac{kT}{eI_E} = \frac{I_C}{I_E} = \alpha_F \tag{9-17}$$

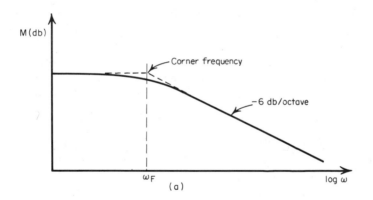

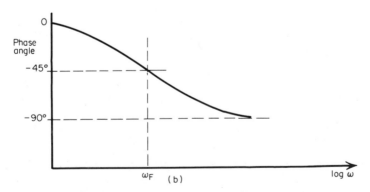

Fig. 9-6 (a) Magnitude and (b) phase versus frequency for single-pole transfer characteristic $1/(1 + j\omega/\omega_F)$.

where α_F = low-frequency value of current gain. Substituting Eq. (9-17) into Eq. (9-16) we obtain

$$i_c = \frac{\alpha_F}{1 + j\omega r_{ib} C_D} \, i_e \qquad (9\text{-}18)$$

The collector current (and hence the effective gain) is seen to decrease as the frequency is increased. The collector current lags behind the emitter current by the phase angle $\tan^{-1} \omega r_{ib} C_D$. The phase angle increases monotonically with frequency and approaches a maximum value of 90° as the frequency approaches infinity. From the definition of diffusion capacitance $C_D = 1/r_{ib}\omega_F$, the effective gain may be expressed as

$$\alpha = \frac{i_c}{i_e} = \frac{\alpha_F}{1 + j\omega/\omega_F} \qquad (9\text{-}19)$$

When the frequency ω equals the forward common-base cutoff frequency ω_F, the magnitude of the gain is $\sqrt{2}/2$ of its low-frequency value and the net phase shift is 45° (Prob. 9-2).

The magnitude and phase characteristics of the term $\alpha/\alpha_F = 1/(1 + j\omega/\omega_F)$ as a function of frequency are sketched in Fig. 9-6. The frequency axes are logarithmic scales and the magnitude M is expressed in db, where

$$M = 20 \log \left|\frac{\alpha}{\alpha_F}\right| \qquad \text{db} \qquad (9\text{-}20)$$

A straight-line asymptotic approximation of the magnitude characteristic is shown in broken lines. For $\omega \ll \omega_F$, $|\alpha/\alpha_F| \approx 1$, and

$$M \approx 20 \log 1 = 0 \qquad (9\text{-}21)$$

Equation (9-21) corresponds to the 0-db straight line of Fig. 9-6a. For $\omega \gg \omega_F$, $|\alpha/\alpha_F| = \omega_F/\omega$, and

$$M = 20 \log \frac{\omega_F}{\omega} = -20 \log \frac{\omega}{\omega_F} \qquad (9\text{-}22)$$

Equation (9-22) corresponds to the sloped line of Fig. 9-6a. For every doubling of frequency (every octave), the magnitude decreases by 6 db. This fact may be shown by considering the difference

$$\begin{aligned} M_{(2\omega)} - M_{(\omega)} &= -20 \log \frac{2\omega}{\omega_F} + 20 \log \frac{\omega}{\omega_F} \\ &= -20 \log 2 - 20 \log \frac{\omega}{\omega_F} + 20 \log \frac{\omega}{\omega_F} \qquad (9\text{-}23) \\ &= -20 \log 2 \approx -6 \text{ db} \end{aligned}$$

The two asymptotes are connected at $\omega = \omega_F$, which point is often called the "corner frequency." The asymptotic method for plotting magnitude response is a convenient means for quickly sketching the approximate characteristics of more complicated amplifier configurations. For example, the magnitude of a transfer characteristic of the form

$$G(\omega) = \frac{A}{[1 + j(\omega/\omega_1)][1 + j(\omega/\omega_2)]} \tag{9-24}$$

can be quickly sketched as shown in Fig. 9-7. There are two corner frequencies, at ω_1 and ω_2, respectively. Starting at the low-frequency end, a horizontal line corresponding to the dc value $20 \log A$ is drawn to the first corner frequency ω_1. From ω_1 to ω_2, the response is reduced 6 db/octave. Beyond ω_2 the second term is also effective, and the response is reduced by 12 db/octave. The phase angle varies by $2 + (-90°) = -180°$ over the frequency range.

The magnitude response for the transfer characteristic

$$G(\omega) = \frac{A(1 + j\omega/\omega_2)}{(1 + j\omega/\omega_1)(1 + j\omega/\omega_3)} \tag{9-25}$$

is shown in Fig. 9-8. At the corner frequency $\omega = \omega_2$, $+6$ db/octave is added to the response since the term appears in the numerator. This term also contributes a positive phase shift of from 0 to 90° so that the total phase shift at $\omega \gg \omega_3$ is $2 \times (-90°) + 90° = 90°$. We will return to these plots when amplifier compensation is discussed later in the chapter.

The frequency-response characteristics for the common-emitter configuration (Fig. 9-4b) may be developed from a simplified equivalent circuit analogous to Fig. 9-5. In particular, it may be shown (Prob. 9-3) that the common-emitter current gain β can be expressed as

$$\beta = \frac{\beta_F}{1 + j(\omega/\omega_F')} \tag{9-26}$$

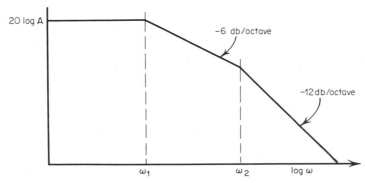

Fig. 9-7   *Asymptotic approximation of two-pole transfer characteristic $A/[(1 + j\omega/\omega_1)(1 + j\omega/\omega_2)]$.*

where $\omega_F' = \omega_F/(\beta_F + 1)$

β_F = low-frequency value of β

Single-transistor Amplifier Limitations. The common-emitter amplifier is the most popular single-transistor amplifier in conventional-circuit design practice. Driving a resistance load R_L, it has a current gain nearly equal to β, a voltage gain equal to $\beta R_L/(r_{bb} + r_{ie})$, and a moderately high input impedance. It has the highest power gain of any single-transistor amplifier configuration. Because of the useful overall characteristics, it also sees application in integrated-circuit designs. Here, however, the configuration possesses the disadvantage of temperature instability of operating point. Variations in operating point would be propagated and amplified by the succeeding direct-coupled stages. Temperature compensation must be applied or another configuration sought.

The temperature instabilities result primarily from the following effects:

1. Leakage current increases with temperature at a rate proportional to $T^{5/2} \exp(-eE_g/kT)$, where E_g is the energy-gap potential ≈ 1.1 ev. A rule of thumb to remember is that in the neighborhood of room temperature, leakage current doubles for every increase of $10°C$ in temperature. The changes in leakage current result in corresponding changes in ohmic-potential drop through biasing resistors, and act as equivalent voltage inputs.

2. For voltage-driven transistors, the collector current increases at a rate proportional to $T^{5/2} \exp[(V_{BE} - E_g)/kT]$.

3. Semiconductor resistors change value with temperature. They can either increase or decrease in value depending upon the doping level and the temperature.

A configuration that avoids most of the deleterious effects of the above is the differential amplifier, which is discussed next. This configuration is the primary one employed in linear integrated circuits.

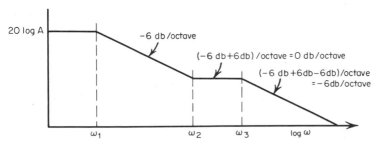

Fig. 9-8 Asymptotic approximation of two-pole transfer characteristic $A(1 + j\omega/\omega_2)/(1 + j\omega/\omega_1)(1 + j\omega/\omega_3)$.

Differential Amplifier

Advantage. The differential amplifier may be described as two identical emitter-coupled amplifiers which interreact to amplify only the difference between their respective input voltages. A commonly employed configuration is shown in Fig. 9-9. The output voltage of the differential amplifier V_{out} is obtained between the two taps, labeled V_{out_1} and V_{out_2}. Changes in V_{in_1} and V_{in_2} in the same direction have relatively little effect on the output voltage.

Consider the advantage of this characteristic. If each of the three temperature effects itemized in the last section occur equally in symmetrically placed components, little change will result, and good temperature stability will be maintained. The key to the problem is to assure that identical changes will result, viz., symmetrically placed components must be as identical as possible in values and in temperature behavior. Further, symmetrically placed components must have the same temperature environment.

In conventional design, the above-mentioned criteria may be met only with difficulty and expense. Transistors must be matched in characteristics (particularly V_{BE}) and packaged together to prevent temperature differentials.

The differential amplifier is a natural circuit, however, for integrated-circuit technology. One may exploit here to the fullest the excellent matching of transistor characteristics and their physically (and therefore thermally) close environment. Significant advantage may also be obtained from good control over resistance ratios and their excellent tracking with temperature.

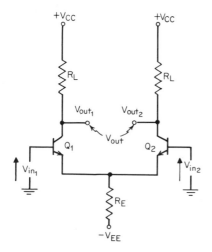

Fig. 9-9 Differential amplifier.

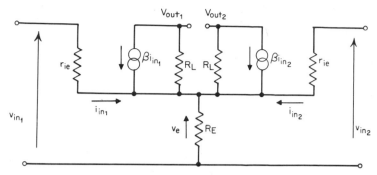

Fig. 9-10 *Differential-amplifier ac-equivalent circuit.*

Characteristics. The low-frequency ac voltage gain and input resistance for the differential amplifier may be determined from the equivalent circuit of Fig. 9-10. The ac (differential) voltage gain is defined as

$$A_D = \frac{v_{out_2} - v_{out_1}}{v_{in_2} - v_{in_1}} \tag{9-27}$$

We write

$$v_{out_2} - v_{out_1} = -\beta R_L (i_{in_2} - i_{in_1}) \tag{9-28}$$

But

$$i_{in_1} = \frac{v_{in_1} - v_e}{r_{ie}}$$
$$i_{in_2} = \frac{v_{in_2} - v_e}{r_{ie}} \tag{9-29}$$

Substituting Eqs. (9-29) into (9-28), we obtain

$$v_{out_2} - v_{out_1} = -\frac{\beta R_L}{r_{ie}} (v_{in_2} - v_{in_1})$$

so that

$$A_D = -\frac{\beta R_L}{r_{ie}} \tag{9-30}$$

The voltage gain is the same as that of the single-transistor common-emitter amplifier.

Now assume that we apply a voltage generator between the inputs so that $v_{in} = v_{in_2} - v_{in_1}$ and $i_{in} = i_{in_2} = -i_{in_1}$. The differential input impedance is defined as

$$z_D = \frac{v_{in}}{i_{in}} = \frac{v_{in_2} - v_{in_1}}{i_{in}} \tag{9-31}$$

From Eqs. (9-29),

$$v_{\text{in}_2} - v_{\text{in}_1} = r_{ie}(i_{\text{in}_2} - i_{\text{in}_1}) = 2r_{ie}i_{\text{in}} \tag{9-32}$$

Therefore

$$z_D = 2r_{ie} \tag{9-33}$$

and we obtain the result that the differential input-impedance is twice that of the common-emitter amplifier.

The common-mode voltage gain is defined as the ratio of the output-difference voltage to a common increase in input voltage (the two input terminals shorted together). For the perfectly matched circuit of Fig. 9-10, it is zero. In practice, some degree of mismatch will result in a finite common-mode gain. A figure of merit for the differential amplifier is the ratio of the differential gain to the common-mode gain, and it is called the common-mode rejection ratio (CMRR). CMRRs varying from 60 to 100 db (1,000 to 100,000) are typical of integrated-circuit amplifiers.

For a given mismatch, the CMRR may be improved by increasing R_E. The dynamic resistance may be made very large by replacing R_E with a constant current source. In practice, a configuration such as that shown in Fig. 9-11 is often employed in integrated circuits.

Linear Integrated-circuit Techniques

The differential amplifier just discussed is a good example of the manner in which certain characteristics of integrated-circuit technology may be

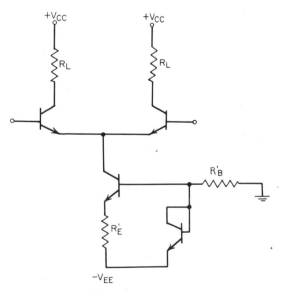

Fig. 9-11 *Differential amplifier using constant current-emitter source.*

employed to advantage in order to circumvent the limitations of other characteristics. As a general rule, it is advantageous to make use of the ratio between parameters rather than their absolute values. Further advantage may be gained through liberal use of transistors and diodes which are high in quality and low in price. We now consider other techniques which exemplify linear integrated-circuit design practice.

Transistor Temperature Compensation.   p-n junction currents are heavily dependent upon temperature. This dependence arises from the thermal generation of charge carriers and the effects of temperature upon physical parameters, such as the diffusion constant. The temperature dependence of collector current may be shown to be approximately

$$I_C \propto T^{5/2} \exp \frac{-e(E_g - V_{BE})}{kT} \tag{9-34}$$

Since increases in temperature will increase both the 5/2 power term and the exponential term, collector current at constant V_{BE} is unstable in a changing temperature environment.

The question arises as to how V_{BE} can be changed with temperature in order to maintain a constant I_C. If Eq. (9-34) is differentiated with respect to temperature, and dI_C/dT set at zero, one obtains

$$\frac{dV_{BE}}{dT} = -\left(\frac{E_g - V_{BE}}{T} + \frac{5}{2}\frac{k}{e}\right) \tag{9-35}$$

At room temperature with $V_{BE} \approx 0.6$ volt, Eq. (9-35) gives

$$\frac{dV_{BE}}{dT} \approx -2 \text{ mv/°C} \tag{9-36}$$

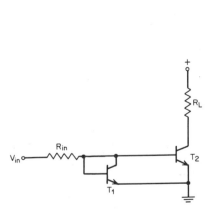

Fig. 9-12   Use of diode-connected transistor for temperature compensation.

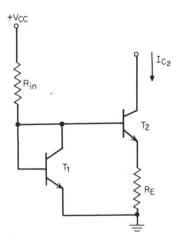

Fig. 9-13   Constant current source.

Thus every centigrade-degree change in temperature is approximately equivalent to a 2-mv change in base-to-emitter voltage. One method* of compensating a transistor for temperature variations is shown in Fig. 9-12. A diode-connected transistor T_1 is used to establish the input voltage to a second transistor T_2. As the temperature is raised, the current through T_1 increases and diverts current from the base of T_2. Compensation is obtained here at the expense of current gain. The current gain is reduced by the ratio of the current through T_1 to the total input current.

Constant Current Source. A variant of the preceding circuit may be used as a current source to generate currents in the microampere range.† Traditional methods require high-value resistors for biasing, which are expensive in integrated-circuit form.

Consider the circuit of Fig. 9-13, where a resistor R_E has been added to the emitter leg of transistor T_2. The base-to-emitter voltage of T_2 is equal to that of T_1 minus the ohmic drop across R_E. Since collector current varies exponentially with V_{BE}, the collector current through T_2 can be made to be significantly smaller than that through T_1 by appropriate selection of R_E.

Modification of Impedance Levels. The amplifying properties of transistors may be used to increase the effective values of resistors and capacitors.

A simple illustration of the multiplication of resistance values is provided by the common-collector amplifier of Fig. 9-14. The effective input resistance of the amplifier is

$$R_{in} = \frac{V_{in}}{I_b}$$

But

$$V_{in} = I_b r_{ie} + I_E R_E$$

$$= I_b \frac{kT}{eI_E} \beta + (1 + \beta)I_b R_E$$

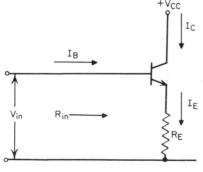

Fig. 9-14   *Common-collector amplifier.*

*Ref. 1.
†Ref. 5.

so that

$$R_{\text{in}} = \frac{kT}{eI_E}\beta + (1+\beta)R_E$$

For $R_E \gg kT/eI_E$,

$$R_{\text{in}} \approx (1+\beta)R_E \qquad\qquad (9\text{-}37)$$

Thus the effective input resistance is approximately R_E times the gain of the transistor.

As a second example, consider the low-pass circuit of Fig. 9-15a. At frequencies sufficiently low with respect to the transistor cutoff frequency, the ac equivalent circuit of Fig. 9-15b may be drawn. The effective gain of the circuit is

$$A_i = \frac{i_L}{i_b} = \frac{\beta i_b[(1/j\omega C_1)/(R_L + 1/j\omega C_1)]}{i_b}$$

$$= \frac{\beta}{1 + j\omega C_1 R_L} \qquad\qquad (9\text{-}38)$$

The cutoff frequency is therefore

$$f_{co} = \frac{1}{2\pi C_1 R_L}$$

The same frequency characteristic can be obtained with a smaller capaci-

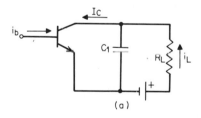

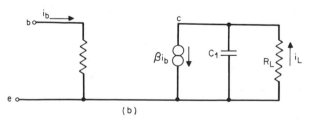

Fig. 9-15 (a) Low-pass circuit and (b) ac equivalent.

tor by using the shunt-feedback arrangement of Fig. 9-16a with the ac equivalent circuit of Fig. 9-16b. The effective gain is

$$A_i = \frac{i_L}{i_{\text{in}}} = \frac{\beta i_b - i_f}{i_f - i_b} \approx \frac{\beta i_b}{i_f + i_b} \tag{9-39}$$

with the assumption that $\beta i_b \gg i_f$. We note that

$$v_{\text{out}} = (\beta i_b - i_f)R_L = i_f z_f - i_b r_{ie}$$

or

$$i_f \approx \frac{i_b(\beta R_L + r_{ie})}{z_f} \approx \frac{i_b \beta R_L}{z_f} \tag{9-40}$$

In arriving at Eq. (9-40), we have used the assumptions $\beta i_b \gg i_f$ and $\beta R_L \gg r_{ie}$. Substituting Eq. (9-40) into Eq. (9-39), we obtain

$$A_i = \frac{\beta}{1 + \beta R_L/z_f} = \frac{\beta}{1 + j\omega\beta C_2 R_L} \tag{9-41}$$

with a cutoff frequency

$$f_{co} = \frac{1}{2\pi\beta C_2 R_L} \tag{9-42}$$

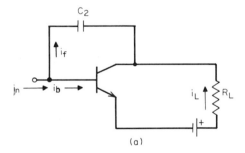

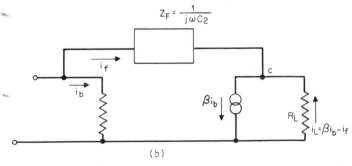

Fig. 9-16 (a) Alternative low-pass circuit and (b) equivalent circuit.

Thus the feedback arrangement of Fig. 9-16 allows the use of a capacitor of factor β smaller than that of Fig. 9-15 to realize the same frequency characteristics.

The preceding simple examples of impedance modification are represent-

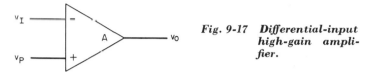

Fig. 9-17 *Differential-input high-gain amplifier.*

ative of a variety of configurations that may be used to limit requirements for high-cost passive components.

Feedback and Compensation

We now consider the practical application of linear integrated circuits in design problems. As a general rule, linear integrated circuits are supplied as differential-input high-gain amplifiers, as shown in Fig. 9-17. The polarities indicated on the input terminals refer to the resulting polarity of the output voltage for a positive input voltage, i.e., v_o is in phase with v_p and 180° out of phase with v_i. If the gain of the amplifier is A, then the output voltage is

$$v_o = A(v_p - v_i) \tag{9-43}$$

In the practical use of the amplifier, it is often necessary to apply negative feedback. Here some fraction of the output voltage is fed back to the negative input terminal. This action decreases the effective gain of the amplifier but has many desirable effects.

The effective gain is then largely determined by the feedback elements rather than the gain of the amplifier (which may vary considerably from device to device). The stability of the gain with parametric drift is improved, the frequency bandwidth is extended, the input impedance is increased and output impedance decreased, and distortion is reduced. Feedback may also have a deleterious effect. At high frequencies, the phase shift through the amplifier may change the character of the feedback from negative to positive. This regenerative effect may cause self-perpetuating high-frequency oscillation to develop, making the amplifier useless for its intended application. The amplifier may be compensated to prevent this effect by adding elements. We now consider the topics of feedback and compensation.

Feedback. The original gain of an amplifier is often referred to as the open-loop gain, and the effective gain after the application of feedback as closed-loop gain. We shall determine the closed-loop gain of the amplifier illustrated in Fig. 9-18. An input voltage v_g is applied to the positive input terminal and some fraction of the output voltage $\beta' v_o$* to the inverting input terminal. From Eq. (9-43) we have

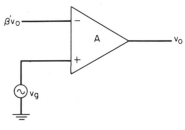

Fig. 9-18 *Basic feedback amplifier.*

$$v_o = A(v_g - \beta' v_o)$$

or

$$v_o = \frac{A v_g}{1 + A\beta'} \qquad (9\text{-}44)$$

The closed-loop gain is therefore

$$G = \frac{v_o}{v_g} = \frac{A}{1 + A\beta'} \qquad (9\text{-}45)$$

The gain is thus reduced by the factor $1 + A\beta'$. We note that for very large A such that $A \gg 1$, Eq. (9-45) reduces to

$$G \approx \frac{1}{\beta'} \qquad (9\text{-}46)$$

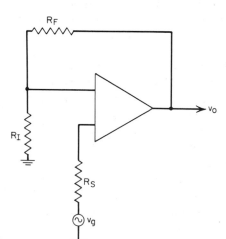

Fig. 9-19 *Feedback arrangement for signal applied to noninverting input.*

*Here β' represents some fraction less than unity and is not to be confused with the common-emitter current gain of a transistor.

Under this condition, the gain is determined by the feedback fraction alone and is independent of the open-loop gain.

In practice, feedback may be accomplished with the voltage divider of Fig. 9-19. If, for simplicity, we neglect the input impedance of the amplifier, we have

$$\beta' = \frac{R_I}{R_I + R_F} \tag{9-47}$$

Resistor R_S is used to equalize the impedance levels as considered from either input terminal. This provides minimum offset, as discussed in the section on the differential amplifier. R_S for this condition is approximately equal to the parallel combination of R_I and R_F:

$$R_S = \frac{R_I R_F}{R_I + R_F} \tag{9-48}$$

If it is desired to apply the input signal to the inverting input, the configuration of Fig. 9-20 may be used. Using linear supposition, we have

$$v_o = -A \left(v_g \frac{R_F}{R_I + R_F} + v_o \frac{R_I}{R_I + R_F} \right)$$

Solving for the closed-loop gain, we obtain

$$G = \frac{-A}{1 + (A + 1)(R_I/R_F)} \approx \frac{R_I}{1 + A(R_I/R_F)} \tag{9-49}$$

so that

$$\beta' = \frac{R_I}{R_F} \tag{9-50}$$

R_S is made equal to the parallel combination of R_I and R_F as in Eq. (9-48).

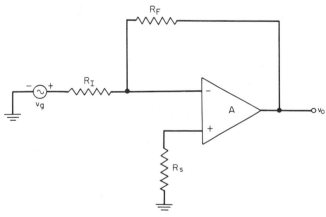

Fig. 9-20 *Feedback arrangement for signal applied to inverting input.*

Finally, a properly balanced feedback arrangement for a double-ended input is shown in Fig. 9-21. The feedback ratio β' is again given by Eq. (9-48).

In addition to these simple amplifier configurations, more complex feedback arrangements may be used to derive a variety of special-purpose circuits, such as integrator, peak detector, Schmitt trigger, etc. Many manufacturers publish helpful guidelines in the form of literature on constructing these circuits.

Compensation. The open-loop gain does not remain constant in frequency but is characterized by a decreasing magnitude and varying phase shift, as illustrated in Fig. 9-22*a* and *b*. The magnitude and phase plots are similar to those discussed earlier in the chapter. The amplifier may be thought to consist of an equivalent frequency-independent amplifier followed by several equivalent RC networks. As each succeeding cutoff (or corner) frequency is reached, the gain magnitude decreases at an additional -6 db/octave rate and the phase undergoes an additional $-90°$ shift. Thus, in Fig. 9-22, we expect a $-90°$ phase shift somewhere in the -6 db/octave region, a $-180°$ phase shift in the -12 db/octave region, etc.

An important consideration is the effect of the phase shift on the feedback. At $-180°$ phase shift, the output voltage has reversed phase and the feedback is now positive. If the loop gain exceeds unity at that point, oscillations result.

The frequency-dependent expression for closed-loop gain is

$$G(\omega) = \frac{A(\omega)}{1 + \beta' A(\omega)} \tag{9-51}$$

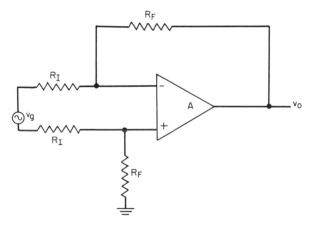

Fig. 9-21 *Feedback arrangement for double-ended input.*

Let the $-180°$ phase point occur at $\omega = \omega_C$, and the magnitude of open-loop gain at this point be A_C. Then

$$A(\omega_C) = -A_C$$

The closed-loop gain at this frequency is

$$G(\omega_C) = \frac{-A_C}{1 - \beta' A_C} \tag{9-52}$$

For stable operation, we must have

$$\beta' < \frac{1}{A_C} \tag{9-53}$$

One method of assuring stable operation for any feedback ratio is to require A_C to be less than unity. In general, we add additional circuit elements to a convenient circuit node if we wish to structure the plots of Fig. 9-21 in order to meet this requirement. Such frequency compensation

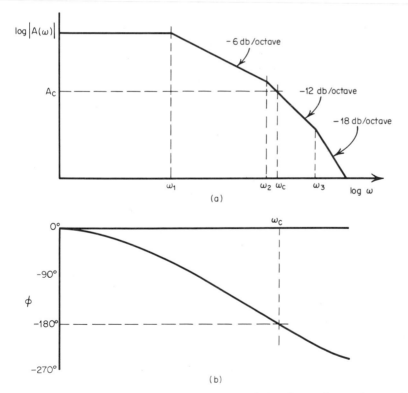

Fig. 9-22   *Variation in (a) magnitude and (b) phase of open-loop gain as a function of frequency.*

may be accomplished internally by the manufacturer, or it may be required as an external accompaniment.

One method of compensation is to provide a low-pass network which reduces the gain below unity prior to the -12 db/octave region. As an example, suppose an amplifier is as shown in Fig. 9-23. The gain at the $-180°$ phase shift is 20 db and therefore the circuit is conditionally unstable. Suppose now that an RC low-pass network is added to the amplifier with a corner frequency at ω' such that the -12 db/octave roll-off does not begin until the gain is unity. To achieve the 60 db attenuation at ω_1 (with a -6 db/octave roll-off), ω' is required to be 10 octaves below ω_1. Thus

$$\omega' = 2^{-10}\omega_1 \approx \omega_1 \times 10^{-3}$$

The required RC time constant is

$$RC = \frac{1}{\omega'} = \frac{1,000}{\omega_1}$$

We note that the above method of compensation considerably degrades the high-frequency response of the amplifier and is suitable only for low-frequency applications. More complex compensation methods exist which do not destroy the frequency response. One such method is to cancel out the pole at $\omega = \omega_2$ by adding a lead network with a corner frequency at ω_2. With the pole at ω_2 eliminated, the response continues to roll off at -6 db/octave until past the 0 db point. A discussion of compensation techniques may be found in Refs. 3 and 4.

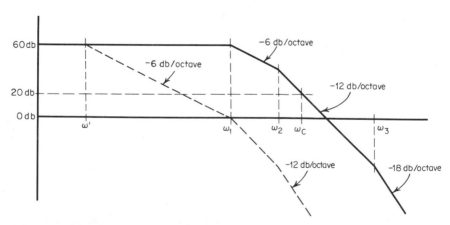

Fig. 9-23 Compensation by adding pole.

REFERENCES

1. Davis, W. R., and H. C. Lin: "Compound Diode-Transistor Structure for Temperature Compensation," *Proc. IEEE*, pp. 1201–1202, September, 1966.
2. Eimbinder, Jerry, ed.: *Linear Integrated Circuits: Theory and Applications*, John Wiley, 1968.
3. Giles, James N.: "Frequency Compensation for an Integrated Operational Amplifier," *Fairchild Semicond. Applic. Bull.* App-117, August, 1965.
4. Radio Corporation of America: *RCA Linear Integrated Circuit Fundamentals*, Technical Series IC-40, Harrison, N. J., 1966.
5. Widlar, R. J.: "Some Circuit Design Techniques for Linear Integrated Circuits," *IEEE Trans. Circuit Theory*, January, 1966.

PROBLEMS

9-1. From Eqs. (9-4), with $V_{CB} = 0$, show that

$$\frac{dI_B}{dV_{BE}} = \frac{e}{kT} \frac{-I_{EO}}{1 - \alpha_F \alpha_I} \exp \frac{eV_{EB}}{kT} \left(1 - \alpha_F + \frac{s}{\omega_F} \right)$$

$$= \frac{1}{r_{ib}} \left(\frac{1}{1 + \beta_F} + \frac{s}{\omega_F} \right)$$

where β_F = low-frequency value of β.

Thus the common-emitter complex imput inpedance is equivalent to the parallel combination of a resistor $r_{ie} = r_{ib} (1 + \beta_F)$ and a capacitor $C_D = 1/r_{ib}\omega_F = (1 + \beta_F)/r_{ie}\omega_F$.

9-2. Using Eq. (9-19), find the magnitude and phase of α/α_F at $\omega = \omega_F$.

9-3. Show that the common-emitter current gain can be expressed as

$$\beta = \frac{\beta_F}{1 + j(\omega/\omega'_F)}$$

where $\omega'_F = \omega_F/(\beta_F + 1)$

β_F = low-frequency value of β

Solutions to Problems

Chapter 2

2-1.

Emitter: $n_0 = N_D = 10^{19}/\text{cm}^3$

$$p_0 = \frac{n_i^2}{n_0} = \frac{2.25 \times 10^{20}}{10^{19}} = 2.25 \times 10^1/\text{cm}^3$$

Base: $p_0 = N_A = 10^{16}/\text{cm}^3$

$$n_0 = \frac{n_i^2}{p_0} = 2.25 \times 10^4/\text{cm}^3$$

Collector: $n_0 = N_D = 10^{15}/\text{cm}^3$

$$p_0 = \frac{n_i^2}{n_0} = 2.25 \times 10^5/\text{cm}^3$$

2-2. The radiation burst acts as an impulse function, $G = 10^9 \delta(t)$. From the differential equation

$$\frac{dn}{dt} = -\frac{n}{\tau}$$

we have

$$\hat{n} = 10^9 \exp\left(-\frac{t}{\tau}\right)$$

For $n = 0.1\, n_0 = 2.25 \times 10^3/\text{cm}^3$, we have

$$t = \ln \frac{10^9}{2.25 \times 10^3} = 1.3 \times 10^{-6} \text{ sec}$$

Since $\hat{p} = \hat{n}$, the percentage of excess majority-carrier concentration is

$$\frac{\hat{p}}{p_0} = \frac{2.25 \times 10^3}{10^{16}} = 2.25 \times 10^{-13} = 2.25 \times 10^{-11} \text{ percent}$$

2-3.

> Emitter: $\rho = 4.6 \times 10^{-4}$ ohm-cm, $R = 0.46$ ohm
>
> Base: $\rho = 1.3$ ohm-cm, $R = 1.3 \times 10^3$ ohms
>
> Collector: $\rho = 4.6$ ohm-cm, $R = 4.6 \times 10^3$ ohms

2-4.

> At emitter edge: $\hat{p}_n = 2.2 \times 10^9/\text{cm}^3$
>
> At collector edge: $\hat{p}_n = -1.0 \times 10^5/\text{cm}^3$

2-5.

$$\alpha = \frac{1}{1 + w^2/2l^2} \qquad \beta = \frac{2l^2}{w^2}$$

To increase gain, use smaller base width w and higher-lifetime ($\tau = l^2/D$) material.

Chapter 3

3-1. The depletion width is expressed in terms of the doping levels and applied voltage by Eq. (2-62):

$$d = \left[\frac{2\epsilon V_t}{e(N_A + N_D)} \right]^{1/2} \left[\left(\frac{N_A}{N_D} \right)^{1/2} + \left(\frac{N_D}{N_A} \right)^{1/2} \right]$$

$$= \left[\frac{2\epsilon V_t (N_D + N_A)}{e N_D N_A} \right]^{1/2}$$

where $V_t = V_B - V_a$. The capacitance is therefore

$$C = \frac{A\epsilon}{d} = \left[\frac{A^2 \epsilon N_D N_A}{2(V_B - V_a)(N_D + N_A)} \right]^{1/2}$$

which is Eq. (3-13).

3-2. *a.* Since

$$\frac{dE}{dx} = \frac{\rho}{\epsilon} = \frac{eax}{\epsilon}$$

$$E = \int \frac{eax}{\epsilon}\, dx = \frac{eax^2}{2\epsilon} + C$$

The constant of integration is evaluated by requiring E to be zero at the edges of the depletion region $+l$:

$$C = -\frac{eal^2}{2\epsilon}$$

Thus

$$E = \frac{ea}{2\epsilon}(x^2 - l^2)$$

b.

$$V_t = -\int_{-l}^{l} E\,dx = -\frac{ea}{2\epsilon}\left[\frac{x^3}{3} - xl^2\right]_{-l}^{l}$$

$$V_t = \frac{2}{3}\frac{l^3\,ea}{\epsilon}$$

c. From part b,

$$l = \left(\frac{3V_t\epsilon}{2ea}\right)^{1/3}$$

$$d = 2l = 2\left(\frac{3V_t\epsilon}{2ea}\right)^{1/3} = \left(\frac{12V_t\epsilon}{ea}\right)^{1/3}$$

d.

$$C = \frac{\epsilon A}{d} = \left(\frac{A^3\epsilon^2 ea}{12V_t\epsilon}\right)^{1/3} = \left(\frac{A^3\epsilon^2 ea}{12V_B}\frac{1}{1 - V_a/V_B}\right)^{1/3}$$

$$= \frac{C_0}{(1 - V_a/V_B)^{1/3}}$$

3-3. The common-emitter gain is, from Eq. (3-40),

$$\beta(\omega) = \frac{H_D}{H_{C_1} + j\omega S_1} = \frac{H_D}{H_{C_1} + j2\pi f S_1}$$

At the unity gain frequency f_T,

$$\left|\beta(f_T)\right| = 1 = \left|\frac{H_D}{H_{C_1} + j2\pi f_T S_1}\right|$$

At this frequency,

$$2\pi f_T S_1 = H_{C_1}$$

and

$$1 = \frac{H_D}{2\pi f_T S_1}$$

Thus

$$f_T = \frac{H_D}{2\pi S_1} = \frac{eAD_p/w}{2\pi(eAw/2)} = \frac{D_p}{\pi w^2}$$

3-4. For $V_G = 3$ or 6 volts, the transistor is in saturation. Equations (3-53) and (3-54) are applicable.

At $V_G = 3$ volts, $I_D = 0.17$ ma, $g_m = 338$ micromhos

At $V_G = 6$ volts, $I_D = 2.7$ ma, $g_m = 1,350$ micromhos

At $V_G = 9$ volts, the transistor is not saturated

The drain current is given by Eq. (3-52), and

$$g_m = \frac{\partial I_D}{\partial V_G} = \frac{\mu_n C_g}{l^2} V_{DS}$$

Using the above, $I_D = 8.4$ ma, $g_m = 2,025$ micromhos.

Chapter 4

4-1. *a.*

$$N = N_0 \operatorname{erfc} \frac{x}{2\sqrt{Dt}}$$

To form a junction in the p-type silicon, a phosphorus concentration of $N = 10^{16}$ is required. Thus

$$\operatorname{erfc} \frac{x}{2\sqrt{Dt}} = \frac{N}{N_0} = \frac{10^{16}}{10^{18}} = 10^{-2}$$

From Fig. 4-2,

$$\frac{x}{2\sqrt{Dt}} = 1.82$$

or

$$t = \frac{1}{D}\left(\frac{x}{2 \times 1.82}\right)^2 = 2.52 \times 10^3 \text{ sec}$$

b. We first determine the phosphorus concentration at 2.5×10^{-4} cm:

$$\frac{x}{2\sqrt{Dt}} = 0.91$$

Hence

$$\operatorname{erfc} \frac{x}{2\sqrt{Dt}} = 0.20$$

and

$$N_p = 0.20\, N_{p_0} = 2.0 \times 10^{17}/\text{cm}^3$$

From this we must subtract the 10^{16} cm³ p-type boron background concentration. The total donor concentration at $x = 2.5 \times 10^{-4}$ cm is therefore $1.9 \times 10^{17}/$cm³; this is the required indium concentration. For the indium,

$$\frac{x}{2\sqrt{Dt}} = 1.49$$

$$\operatorname{erfc} \frac{x}{2\sqrt{Dt}} = 3.4 \times 10^{-2} = \frac{N_I}{N_{I_0}}$$

Therefore

$$N_{I_0} = \frac{N_I}{3.4 \times 10^{-2}} = \frac{19 \times 10^{16}}{3.4 \times 10^{-2}} = 5.3 \times 10^{18}/\text{cm}^3$$

Chapter 5

5-1. *a.* With V_{in} below the threshold voltage, transistor T_1 is biased in cutoff (non-conducting). The gate-to-source voltage of transistor T_2, V_{GS_2}, will find a stable value equal to the threshold voltage in order to support the small measurement current:

$$V_{GS_2} = V_{th}$$

But
$$V_{GS_2} = V_{DD} - V_{\text{out}}$$
Therefore
$$V_{\text{out}} = V_{DD} - V_{th}$$

b. T_2 operates in saturation for all values of V_{in} since
$$V_{DS_2} = V_{GS_2} > V_{GS_2} - V_{th}$$

As V_{th} is increased to V_{th}, T_1 begins to conduct. At the point of conduction, $V_{GS_1} = V_{th}$ and $V_{DS_1} = V_{DD} - V_{th}$ (from part a).
Therefore
$$V_{DS_1} > V_{GS_1} - V_{th} = 0$$

and T_1 is in saturation. T_1 will leave saturation when
$$V_{DS_1} \le V_{GS_1} - V_{th}$$
or
$$V_{\text{out}} \le V_{\text{in}} - V_{th}$$

To solve for this condition, we equate the currents conducted by the two transistors:
$$I_{D_2} = \frac{\mu_n C_2}{2l^2} (V_{GS_2} - V_{th})^2$$
$$= \frac{\mu_n C_2}{2l^2} (V_{DD} - V_{\text{out}} - V_{th})^2$$
$$I_{D_1} = \frac{\mu_n C_1}{2l^2} (V_{GS_1} - V_{th})^2$$
$$= \frac{\mu_n C_1}{2l^2} (V_{\text{in}} - V_{th})^2$$

Setting $I_{D_2} = I_{D_1}$, and taking the physically meaningful positive square roots, we obtain
$$C_2^{1/2} (V_{DD} - V_{\text{out}} - V_{th}) = C_1^{1/2} (V_{\text{in}} - V_{th})$$

To find the input voltage where T_1 leaves saturation, we substitute into the above equation $V_{\text{out}} = V_{\text{in}} - V_{th}$ and find
$$V_{\text{in}} = \frac{(C_2/C_1)^{1/2} V_{DD} + V_{th}}{1 + (C_2/C_1)^{1/2}}$$

Thus
$$V_{th} \le V_{\text{in}} < \frac{(C_2/C_1)^{1/2} V_{DD} + V_{th}}{1 + (C_2/C_1)^{1/2}}$$

for T_1 in saturation.

c. From the equated current expressions,
$$C_2^{1/2} V_{\text{out}} = C_2^{1/2}(V_{DD} - V_{th}) - C_1^{1/2}(V_{\text{in}} - V_{th})$$
we obtain
$$V_{\text{out}} = V_{DD} - V_{th} - \left(\frac{C_1}{C_2}\right)^{1/2} (V_{\text{in}} - V_{th})$$

d. From part b, the range of input voltage for T_1 in saturation is $0.2\,V_{DD} \leq V_{in} < 0.4\,V_{DD}$. From part c,

$$V_{out} = V_{DD} - 0.2V_{DD} - (9)^{1/2}(V_{in} - 0.2V_{DD})$$
$$= 1.4V_{DD} - 3V_{in}$$

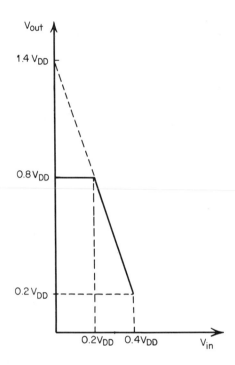

Chapter 6

6-1. Theorem Ia uses IIb, Va, IVb, Vb, IIa.

Theorem Ib:

$$
\begin{array}{ll}
A + 0 = A & \text{(II}a) \\
A + A\bar{A} = A & \text{(V}b) \\
(A + A)\cdot(A + \bar{A}) = A & \text{(IV}a) \\
(A + A)\cdot 1 = A & \text{(V}a) \\
A + A = A & \text{(II}b)
\end{array}
$$

6-2. *a.*

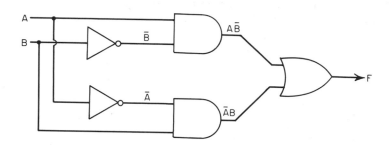

$$F = A\bar{B} + \bar{A}B$$

b.

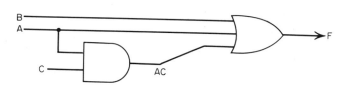

$$F = A + B + AC$$

c.

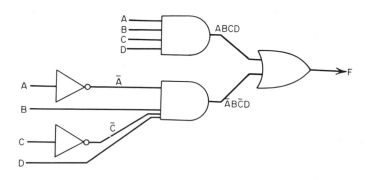

Expression in part *b* can be simplified to $F = A + B$ with a corresponding simplification in block diagram.

6-3.

$$(A \oplus B) \oplus C = \overline{(A \oplus B)}C + (A \oplus B)\bar{C}$$

$$= \overline{(\bar{A}B + A\bar{B})}C + (\bar{A}B + A\bar{B})\bar{C}$$

$$= (\overline{\bar{A}B}) \, (\overline{A\bar{B}})C + \bar{A}B\bar{C} + A\bar{B}\bar{C}$$

$$= (A + \bar{B}) \, (\bar{A} + B)C + \bar{A}B\bar{C} + A\bar{B}\bar{C}$$

$$\overset{0}{} \qquad\qquad \overset{0}{}$$

$$= (A\bar{A} + AB + \bar{A}\bar{B} + B\bar{B})C + \bar{A}B\bar{C} + A\bar{B}\bar{C}$$

$$= \bar{A}B\bar{C} + A\bar{B}\bar{C} + \bar{A}\bar{B}C + ABC$$

$$A \oplus (B \oplus C) = \bar{A}(B \oplus C) + A\overline{(B \oplus C)}$$

$$= \bar{A}(\bar{B}C + B\bar{C}) + A\overline{(\bar{B}C + B\bar{C})}$$

$$= \bar{A}\bar{B}C + \bar{A}B\bar{C} + A(B + \bar{C})(\bar{B} + C)$$

$$= \bar{A}\bar{B}C + \bar{A}B\bar{C} + A(B\bar{B} + BC + \bar{B}\bar{C} + C\bar{C})$$

$$= \bar{A}B\bar{C} + A\bar{B}\bar{C} + \bar{A}\bar{B}C + ABC$$

6-4. *a.*

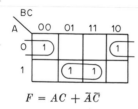

$$F = AC + \bar{A}\bar{C}$$

b.

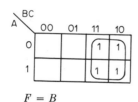

$$F = B$$

c.

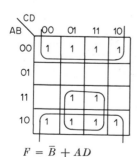

$$F = \bar{B} + AD$$

6-5. Let the output of the circuit, F, be logical 1 when danger exists. We examine all possibilities in the truth table below and simplify by means of a Karnaugh map.

M	D	G	C	F
0	0	0	0	0
0	0	0	1	0
0	0	1	0	0
0	0	1	1	1
0	1	0	0	0
0	1	0	1	0
0	1	1	0	1
0	1	1	1	1
1	0	0	0	1
1	0	0	1	1
1	0	1	0	0
1	0	1	1	0
1	1	0	0	1
1	1	0	1	0
1	1	1	0	0
1	1	1	1	0

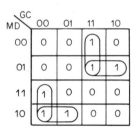

$$F = M\overline{G}\overline{C} + M\overline{D}\overline{G} + \overline{M}DG + \overline{M}GC$$

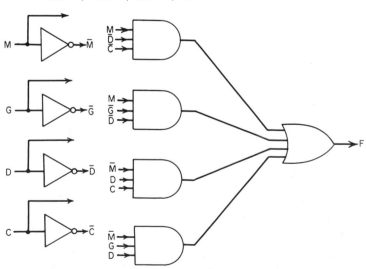

6-6. We assume that the complements of the input variables are available or can be obtained through use of single-input NAND or NOR gates.

a. From map of Prob. 6-5, the reduced expressions are

$$F = M\bar{D}\bar{C} + M\bar{G}\bar{D} + \bar{M}DC + \bar{M}GD$$

$$\bar{F} = \bar{M}\bar{G} + MG + MDC + \bar{M}\bar{D}\bar{C}$$

NAND:

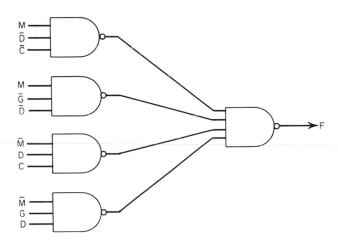

NOR:

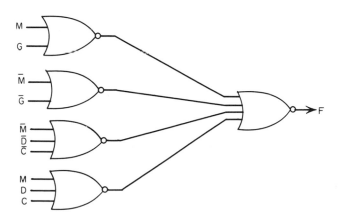

b. NAND: $F = A\bar{B} + \bar{A}B$

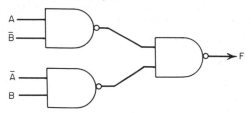

NOR: $F = AB + \bar{A}\bar{B}$

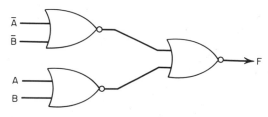

c. NAND: $F = A + B$

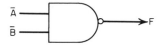

NOR: $F = \bar{A}\bar{B}$

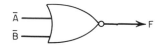

d. NAND: $F = AB$

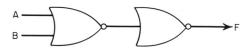

NOR: $\bar{F} = \bar{A} + \bar{B}$

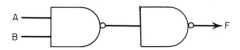

Chapter 7

7-1. State table for problem:

Present state				Next state				Required inputs							
Q_4	Q_3	Q_2	Q_1	Q_4	Q_3	Q_2	Q_1	R_4	S_4	R_3	S_3	R_2	S_2	R_1	S_1
0	0	0	0	0	0	0	1	×	0	×	0	×	0	0	1
0	0	0	1	0	0	1	1	×	0	×	0	0	1	0	×
0	0	1	1	0	0	1	0	×	0	×	0	0	×	1	0
0	0	1	0	0	1	1	0	×	0	0	1	0	×	×	0
0	1	1	0	0	1	1	1	×	0	0	×	0	×	0	1
0	1	1	1	0	1	0	1	×	0	0	×	1	0	0	×
0	1	0	1	1	1	0	1	0	1	0	×	×	0	0	×
1	1	0	1	1	1	0	0	0	×	0	×	×	0	1	0
1	1	0	0	0	1	0	0	1	0	0	×	×	0	×	0
0	1	0	0	0	0	0	0	×	0	1	0	×	0	×	0

Karnaugh maps for the inputs:

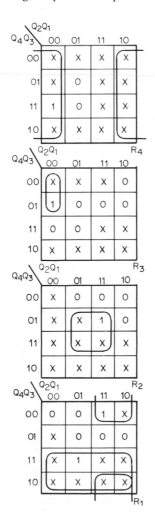

Input equations:

$$R_4 = \bar{Q}_1$$
$$S_4 = Q_3\bar{Q}_2 Q_1$$
$$R_3 = \bar{Q}_4\bar{Q}_2\bar{Q}_1$$
$$S_3 = Q_2\bar{Q}_1$$
$$R_2 = Q_3 Q_1$$
$$S_2 = \bar{Q}_3 Q_1$$
$$R_1 = Q_4 + \bar{Q}_3 Q_2$$
$$S_1 = \bar{Q}_3\bar{Q}_2 + Q_3 Q_2$$

7-2. State table for problem:

Present state				Next state				T flip-flop prob.				D flip-flop prob.			
Q_4	Q_3	Q_2	Q_1	Q_4	Q_3	Q_2	Q_1	T_4	T_3	T_2	T_1	D_4	D_3	D_2	D_1
0	0	0	0	0	0	0	1	0	0	0	1	0	0	0	1
0	0	0	1	0	0	1	1	0	0	1	0	0	0	1	1
0	0	1	1	0	0	1	0	0	0	0	1	0	0	1	0
0	0	1	0	0	1	1	0	0	1	0	0	0	1	1	0
0	1	1	0	0	1	1	1	0	0	0	1	0	1	1	1
0	1	1	1	0	1	0	1	0	0	1	0	0	1	0	1
0	1	0	1	1	1	0	1	1	0	0	0	1	1	0	1
1	1	0	1	1	1	0	0	0	0	0	1	1	1	0	0
1	1	0	0	0	1	0	0	1	0	0	0	0	1	0	0
0	1	0	0	0	0	0	0	0	1	0	0	0	0	0	0

T flip-flop problem:

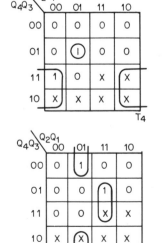

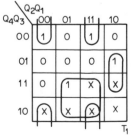

D flip-flop problem:

Q_4Q_3 \ Q_2Q_1	00	01	11	10
00	0	0	0	0
01	0	1	0	0
11	0	1	X	X
10	X	X	X	X

D_4

Q_4Q_3 \ Q_2Q_1	00	01	11	10
00	0	0	0	1
01	0	1	1	1
11	1	1	X	X
10	X	X	X	X

D_3

Q_4Q_3 \ Q_2Q_1	00	01	11	10
00	0	1	1	1
01	0	0	0	1
11	0	0	X	X
10	X	X	X	X

D_2

Q_4Q_3 \ Q_2Q_1	00	0	11	10
00	1	1	0	0
01	0	1	1	1
11	0	0	X	X
10	X	X	X	X

D_1

Input equations:

T flip-flop problem:

$$T_4 = \bar{Q}_4 Q_3 \bar{Q}_2 Q_1 + Q_4 \bar{Q}_1$$
$$T_3 = \bar{Q}_4 Q_3 \bar{Q}_2 \bar{Q}_1 + \bar{Q}_3 Q_2 \bar{Q}_1$$
$$T_2 = \bar{Q}_3 \bar{Q}_2 Q_1 + Q_3 Q_2 Q_1$$
$$T_1 = \bar{Q}_3 \bar{Q}_2 \bar{Q}_1 + \bar{Q}_3 Q_2 Q_1 + Q_3 Q_2 \bar{Q}_1 + Q_4 Q_1$$

D flip-flop problem:

$$D_4 = Q_3 \bar{Q}_2 Q_1$$
$$D_3 = Q_3 Q_1 + Q_2 \bar{Q}_1 + Q_4$$
$$D_2 = \bar{Q}_3 Q_1 + Q_2 \bar{Q}_1$$
$$D_1 = \bar{Q}_4 Q_3 Q_1 + Q_3 Q_2 + \bar{Q}_3 Q_2$$

7-3. State table:

Present state	Next state Input (X)		Output (Z)
	0	1	
a	a	b	0
b	a	c	0
c	d	a	0
d	e	a	0
e	a	f	0
f	g	a	0
g	h	h	1
h	a	a	1

State assignment: A minimum of three flip-flops is required. The following assignment is arbitrarily selected.

State	Q_3	Q_2	Q_1	State	Q_3	Q_2	Q_1
a	0	0	0	e	1	1	0
b	0	0	1	f	1	1	1
c	0	1	1	g	1	0	1
d	0	1	0	h	1	0	0

State table with assignment:

Present state			Next state $(Q_3Q_2Q_1)$ Input						Output (Z)
Q_3	Q_2	Q_1		0			1		
0	0	0	0	0	0	0	0	1	0
0	0	1	0	0	0	0	1	1	0
0	1	1	0	1	0	0	0	0	0
0	1	0	1	1	0	0	0	0	0
1	1	0	0	0	0	1	1	1	0
1	1	1	1	0	1	0	0	0	0
1	0	1	1	0	0	1	0	0	1
1	0	0	0	0	0	0	0	0	1

The problem is now completely specified. For the design we will employ D-type flip-flops. The input requirements for the flip-flops as a function of the present states and the system input X are listed next.

Present state				Next state			Input requirements			System output
Q_3	Q_2	Q_1	X	Q_3	Q_2	Q_1	D_3	D_2	D_1	Z
0	0	0	0	0	0	0	0	0	0	0
0	0	0	1	0	0	1	0	0	1	
0	0	1	0	0	0	0	0	0	0	0
0	0	1	1	0	1	1	0	1	1	
0	1	1	0	0	1	0	0	1	0	0
0	1	1	1	0	0	0	0	0	0	
0	1	0	0	1	1	0	1	1	0	0
0	1	0	1	0	0	0	0	0	0	
1	1	0	0	0	0	0	0	0	0	0
1	1	0	1	1	1	1	1	1	1	
1	1	1	0	1	0	1	1	0	1	0
1	1	1	1	0	0	0	0	0	0	
1	0	1	0	1	0	0	1	0	0	1
1	0	1	1	1	0	0	1	0	0	
1	0	0	0	0	0	0	0	0	0	1
1	0	0	1	0	0	0	0	0	0	

Equations for the inputs and the system output are determined from the Karnaugh maps below.

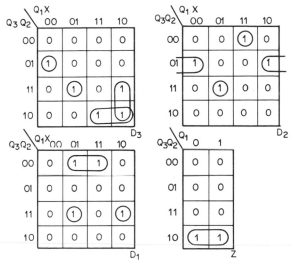

$$D_3 = Q_3Q_1\bar{X} + Q_3\bar{Q}_2Q_1 + \bar{Q}_3Q_2\bar{Q}_1\bar{X} + Q_3Q_2\bar{Q}_1X$$
$$D_2 = \bar{Q}_3Q_2\bar{X} + \bar{Q}_3\bar{Q}_2Q_1X + Q_3Q_2\bar{Q}_1X$$
$$D_1 = \bar{Q}_3\bar{Q}_2X + Q_3Q_2\bar{Q}_1X + Q_3Q_2Q_1\bar{X}$$
$$Z \;= Q_3\bar{Q}_2$$

7-4. A variety of designs may be proposed. One possible design employs a mod 8 counter, a mod 5 counter, a selection network, and miscellaneous gates and DC *RS* flip-flops, as shown in the figure.

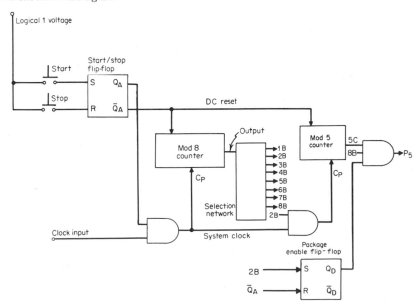

The states of the mod 8 counter, as dictated by the selection network, perform the following functions:

$1B$: Initiate "cut" operation.

$2B$: Initiate "rough sharpen" operation. Enable mod 4 counter to be advanced. Set package-enable flip-flop.

$3B$: Initiate "fine sharpen" operation.

$4B$: Initiate "rough strop" operation.

$5B$: Initiate "fine strop" operation.

$6B$: Initiate "print trademark" operation.

$7B$: Initiate "store in shute" operation.

$8B$: Initiate packaging operation if mod 5 counter is in fifth state ($5C$).

The mod 5 counter, which is advanced one count for each complete cycle of the mod 8 counter, enables the packaging operation ($P5$) when 5 blades have been stored in the shute. The initial states of the counters are $8B$ and $5C$, and are reset to these states when the start-stop flip-flop is reset. Since the initial states are those which initiate the packaging operation, provision is made (via the package-enable flip-flop) to inhibit the packaging operation when the system is initially started. The package-enable flip-flop is set during the first cycle of the mod 8 counter, and remains set until the system is stopped via the start-stop flip-flop. Finally, the start-stop flip-flop resets the counters and the package-enable flip-flop when reset, and enables the clock input to the system when set. The momentary-contact start-and-stop switches set and reset the start-stop flip-flop.

Mod 8 counter state assignment:

State	Q_{3B}	Q_{2B}	Q_{1B}
$1B$	0	0	1
$2B$	0	1	0
$3B$	0	1	1
$4B$	1	0	0
$5B$	1	0	1
$6B$	1	1	0
$7B$	1	1	1
$8B$	0	0	0

Input equations:

$$J_{3B} = K_{3B} = Q_2 Q_1$$
$$J_{2B} = K_{2B} = Q_1$$
$$J_{1B} = K_{1B} = 1$$

DC reset

$$R_{3B} = R_{2B} = R_{1B} = \bar{Q}_A$$
$$S_{3B} = S_{2B} = S_{1B} = 0$$

Mod 5 counter state assignment:

State	Q_{3B}	Q_{2B}	Q_{1B}
1C	0	0	0
2C	0	0	1
3C	0	1	0
4C	0	1	1
5C	1	1	1

Input equations:

$$J_{3C} = Q_2 Q_1$$
$$K_{3C} = 1$$
$$J_{2C} = Q_1$$
$$K_{2C} = Q_3$$
$$J_{1C} = 1$$
$$K_{1C} = Q_3 + \bar{Q}_2$$

DC reset:

$$R_{3C} = R_{2C} = R_{1C} = 0$$
$$S_{3C} = S_{2C} = S_{1C} = \bar{Q}_A$$

Chapter 8

8-1. The currents through T_1 and T_2 are

$$I_{D_2} = \frac{\mu_n C_2}{2l^2} (V_{GS_2} - V_{th})^2$$

$$= \frac{\mu_n C_2}{2l^2} (V_{DD} - V_{out} - V_{th})^2$$

$$I_{D_1} = \frac{\mu_n C_1}{l^2} \left[(V_{GS_1} - V_{th}) V_{DS_1} - \frac{V_{DS_1}^2}{2} \right]$$

$$= \frac{\mu_n C_1}{l^2} \left[(V_{DD} - V_{th}) V_{out} - \frac{V_{out}^2}{2} \right]$$

Equate the two currents to obtain

$$\frac{C_2}{2}(V_{DD} - V_{out} - V_{th})^2 = C_1 \left[(V_{DD} - V_{th}) V_{out} - \frac{V_{out}^2}{2} \right]$$

Solve for V_{out} (taking the physically meaningful root) to get

$$V_{out} = (V_{DD} - V_{th}) \left[1 - \left(\frac{C_1}{C_1 + C_2} \right)^{1/2} \right]$$

Since gate capacitance is proportional to channel width, Eq. (8-5) follows directly.

Chapter 9

9-1. Setting $V_{CB} = 0$ in Eqs. (9-4) and noting that $I_B = -I_C - I_E$, we have

$$I_B = \frac{I_{EO}}{1 - \alpha_F \alpha_I}\left(1 - \alpha_F + \frac{s}{\omega_F}\right)\exp\frac{eV_{EB}}{kT}$$

$$\frac{dI_B}{dV_{BE}} = -\frac{dI_B}{dV_{EB}}$$

$$= \left(\frac{e}{kT}\frac{-I_{EO}}{1 - \alpha_F \alpha_I}\exp\frac{eV_{EB}}{kT}\right)\left(1 - \alpha_F + \frac{s}{\omega_F}\right)$$

9-2.

$$\frac{\alpha}{\alpha_F} = \frac{1}{1 + j(\omega/\omega_F)}$$

For $\omega = \omega_F$,

$$\frac{\alpha}{\alpha_F} = \frac{1}{1 + j} = \frac{1}{2} - j\frac{1}{2}$$

The magnitude is

$$\left|\frac{\alpha}{\alpha_F}\right| = \sqrt{\left(\frac{1}{2}\right)^2 + \left(-\frac{1}{2}\right)^2} = \frac{\sqrt{2}}{2}$$

The phase is

$$\phi = \tan^{-1}\left(\frac{-1/2}{1/2}\right) = -45°$$

9-3. From the results of Prob. 9-1,

$$v_{be} = \frac{i_b}{1/r_{ie} + j\omega C_D}$$

then

$$i_c = -g_m v_{be} = \frac{-g_m r_{ie}}{1 + j\omega r_{ie}C_D}$$

$$= \frac{\beta_F}{1 + j\omega r_{ie}C_D} = \frac{\beta_F}{1 + j(\omega/\omega'_F)}$$

where $\beta_F = -g_m r_{ie}$ = low-frequency common-emitter current gain
$\omega'_F = 1/r_{ie}C_D = 1/[(r_{ie})(1 + \beta_F)/r_{ie}\omega_F] = \omega_F/1 + \beta_F$

Index